Natural Environmental Change

As this book was written, the world changed, as it has done so for all of the 5 billion years of Earth history. The dynamism of planet Earth is especially well illustrated by the transformation that has occurred during the last 3 million years. During this time there have been numerous cold and warm episodes, as global temperatures have fallen by as much as 10°C and then risen again. Today, researchers are probing archives of environmental data for information on past change, in order to help predict, and hence manage, future change caused by both natural and human agencies.

Following two introductory chapters, which examine theories about environmental change and provide a summary of Earth history, there are four chapters on the record of environmental change as revealed by data from various archives. These archives include ocean sediments, ice cores, terrestrial deposits (such as glacial moraines and lake sediments), tree rings, and historical and meteorological records. The subsequent three chapters detail the changes that have occurred in high, middle and low latitudes, while the conclusion offers a perspective on the strengths and weaknesses of current understanding.

Natural Environmental Change offers a concise introduction to a stimulating topic of interest and relevance to students pursuing courses in geography, Earth sciences and environmental sciences. Each chapter provides a broad spectrum of case studies and further reading and there is an extensive bibliography.

A.M. Mannion is a Senior Lecturer in Geography at the University of Reading.

Routledge Introductions to Environment Series

Published and Forthcoming Titles

Titles under Series Editors:
Rita Gardner and A.M. Mannion

Environmental Science texts

Environmental Biology
Environmental Chemistry and Physics
Environmental Geology
Environmental Engineering
Environmental Archaeology
Atmospheric Systems
Hydrological Systems
Oceanic Systems
Coastal Systems
Fluvial Systems
Soil Systems
Glacial Systems
Ecosystems
Landscape Systems
Natural Environmental Change

Titles under Series Editors:
David Pepper and Phil O'Keefe:

Environment and Society texts

Environment and Economics
Environment and Politics
Environment and Law
Environment and Philosophy
Environment and Planning
Environment and Social Theory
Environment and Political Theory
Business and Environment

Key Environmental Topics texts

Biodiversity and Conservation
Environmental Hazards
Natural Environmental Change
Environmental Monitoring
Climate Change
Land Use and Abuse
Water Resources
Pollution
Waste and the Environment
Energy Resources
Agriculture
Wetland Environments
Energy, Society and Environment

Environmental Sustainability
Gender and Environment
Environment and Society
Tourism and Environment
Environmental Management
Environmental Values
Representations of the Environment
Environment and Health
Environmental Movements
History of Environmental Ideas
Environment and Technology
Environment and the City
Case Studies for Environmental Studies

Routledge Introductions to Environment

Natural Environmental Change

The last 3 million years

A.M. Mannion

London and New York

First published 1999
by Routledge
11 New Fetter Lane, London EC4P 4EE

Simultaneously published in the USA and Canada
by Routledge
29 West 35th Street, New York, NY 10001

Typeset in Times by RefineCatch Ltd, Bungay, Suffolk
Printed and bound in Great Britain by
TJ International Ltd, Padstow, Cornwall

British Library Cataloguing in Publication Data
A catalogue record for this book is available from the British Library

Library of Congress Cataloging in Publication Data
Mannion, Antoinette M.
    Natural environmental change / Antoinette M. Mannion.
        p. cm. – (Routledge introductions to environment series)
    Includes bibliographical references.
    1. Global environmental change. 2. Climatic changes. I. Title.
II. Series.
GE149.M37   1999   98–25877
363.7—dc21

ISBN 0–415–13932–5 (hbk)
ISBN 0–415–13933–3 (pbk)

For MDT

Habendum et tenendum

Πάντα ρεῖ, οὐδέν μένει
(All is flux, nothing is stationary)

Heracleitus, philosopher of ancient Greece
c. 535–*c.* 475 BC

Contents

Series editors' preface   *ix*

List of figures   *xi*

List of tables   *xiv*

List of boxes   *xv*

Author's preface   *xvi*

Acknowledgements   *xvii*

Chapter 1   Introduction   1

Chapter 2   Natural environmental change: the long-term geological
             record   15

Chapter 3   The record of environmental change in ocean sediments   33

Chapter 4   The record of environmental change in ice cores   50

Chapter 5   The record of environmental change in continental archives   64

Chapter 6   The record of environmental change in tree rings and
             historical and meteorological records   90

Chapter 7   Environmental change in high latitudes   106

Chapter 8   Environmental change in middle latitudes   123

Chapter 9   Environmental change in low latitudes   147

Chapter 10   Conclusions   164

Glossary   172

Bibliography   177

Index   193

Series editors' preface
Environmental Science titles

The last few years have witnessed tremendous changes in the syllabi of environmentally related courses at Advanced Level and in tertiary education. Moreover, there have been major alterations in the way degrees and diploma courses are organised in colleges and universities. Syllabus changes reflect the increasing interest in environmental issues, their significance in a political context and their increasing relevance in everyday life. Consequently, the 'environment' has become a focus not only in courses traditionally concerned with geography, environmental science and ecology but also in agriculture, economics, politics, law, sociology, chemistry, physics, biology and philosophy. Simultaneously, changes in course organisation have occurred in order to facilitate both generalisation and specialisation; increasing flexibility within and between institutions is encouraging diversification and especially the facilitation of teaching via modularisation. The latter involves the compartmentalisation of information, which is presented in short, concentrated courses that, on the one hand, are self-contained but which, on the other hand, are related to prerequisite, parallel, and/or advanced modules.

These innovations in curricula and their organisation have caused teachers, academics and publishers to reappraise the style and content of published works. Whilst many traditionally styled texts dealing with a well-defined discipline, e.g. physical geography or ecology, remain apposite, there is a mounting demand for short, concise and specifically focused texts suitable for modular degree/diploma courses. In order to accommodate these needs, Routledge have devised the Introduction to Environment Series, which comprises Environmental Science and Environmental Studies. The former broadly encompasses subject matter that pertains to the nature and operation of the environment, and the latter concerns the human dimension as a dominant force within, and a recipient of, environmental processes and change. Although this distinction is made, it is purely arbitrary and is made for practical rather than theoretical purposes; it does not deny the holistic nature of the environment and its all-pervading significance. Indeed, every effort has been made by authors to refer to such interrelationships and to provide information to expedite further study.

This series is intended to fire the enthusiasm of students and their teachers/

lecturers. Each text is well illustrated and numerous case studies are provided to underpin general theory. Further reading is also furnished to assist those who wish to reinforce and extend their studies. The authors, editors and publishers have made every effort to provide a series of exciting and innovative texts that will not only offer invaluable learning resources and supply a teaching manual, but also act as a source of inspiration.

A.M. Mannion and Rita Gardner
1997

Series International Advisory Board

Australasia: Dr P. Curson and Dr P. Mitchell, Macquarie University

North America: Professor L. Lewis, Clark University; Professor L. Rubinoff, Trent University

Europe: Professor P. Glasbergen, University of Utrecht; Professor van Dam-Mieras, Open University, The Netherlands

Notes on the text

Bold is used in the text to denote words defined in the Glossary.

Unless stated otherwise, all depressions or increases in temperature are reported on a mean annual basis.

K years BP = 10^3 years before present.

Figures

1.1	The relationships between the lithosphere, hydrosphere, atmosphere and biosphere	1
1.2	The succession of European ice ages as determined by Penck and Brückner in 1909	5
1.3	William Morris Davis' cycle of erosion	8
1.4	The major components and interrelationships in ecosystems	10
2.1	The structure of the Earth	16
2.2	The major environmental changes that have occurred throughout Earth history	19
2.3	Periods of biotic extinction throughout geological time	23
2.4	Correlations between various Quaternary sequences in the Northern Hemisphere	26–7
3.1	An oxygen isotope record for the last 2.6×10^6 years	36
3.2	The relationships between oxygen isotope stratigraphy, terminations, and alkenone stratigraphy	37
3.3	Examples of foraminifera (greatly magnified)	38
3.4	Examples of diatoms (greatly magnified)	40
3.5	Examples of radiolarians (greatly magnified)	41
3.6	Examples of coccolithophores (greatly magnified)	42
3.7	The reconstruction of sea-surface temperatures in August for the last glacial maximum *c.* 18K years BP (A) in relation to those of the present day (B)	48
4.1	The location of major ice-core sites	51
4.2	The oxygen isotope record of the Vostok ice core, Antarctica	52
4.3	The oxygen isotope records of the GRIP and GISP2 (Greenland) ice cores	53
4.4	The deuterium record from the Vostok ice core, Antarctica	53
4.5	The carbon dioxide and methane records from the Vostok ice core, Antarctica and the methane record from the GISP2 and GRIP (Greenland) ice cores	54
4.6	The concentrations of sodium and aluminium in the Vostok ice core, Antarctica	55
4.7	The record of beryllium (^{10}Be) from the Vostok ice core, Antarctica	56

4.8 The acidity profile of the Vostok ice core, Antarctica   57

4.9 The record of methanesulphonic acid from the Vostok ice core,
Antarctica   58

4.10 Reconstructed temperature changes for the last *c.* 18K years
from the GISP2 (Greenland) ice core   61

5.1 Types of glacial deposits   65

5.2 The maximum extent of ice cover in Britain during the last three
major glaciations   67

5.3 (A) The major ice limits during the last (Wisconsin) ice advance
in North America. (B) The southern margins of the last three
major ice advances in Europe   68

5.4 Maximum extent of ice sheets and permafrost at the last glacial
maximum   69

5.5 Locations of long lake sequences referred to in the text   71

5.6 Correlations between four long lake sequences and the marine
oxygen isotope record   72

5.7 Land-cover types during the Eemian interglacial   75

5.8 Pollen diagram from Hockham Mere, East Anglia   78–9

5.9 The major forest types *c.* 5.5K years BP   80

5.10 The Baoji loess (200 km west of Xian) grain-size record in
relation to the oxygen isotope record of North Atlantic sediment
core DSDP 607   83

6.1 Dendroclimatological data from German sub-fossil tree remains
in relation to oxygen isotope data from the GRIP ice core   94

6.2 Reconstruction of the departure of summer temperatures from
average values (December to March) from a south-central Chile
tree-ring record for the last 3.5K years   94

6.3 (A) The dendroclimatic record from the Tarvagatay Mountains,
Mongolia. (B) Reconstructed summer temperatures from
variations in tree-ring sequences in the northern Urals, Siberia   96

6.4 Dry and very dry indices (DVDI) of climate in China, as derived
from historical records   98

6.5 The record of drought in Crete 1547/48 to 1644/45, as
reconstructed from documentary evidence   99

6.6 Mean annual temperature anomalies for the Northern and
Southern hemispheres, 1950–present   103

7.1 The Arctic region today and the location of sites referred to in
the text   108

7.2 A summary of the stratigraphy and the palaeoenvironmental
inferences from sediment cores extracted from Northward Ridge,
Amerasia Basin   109

8.1 Middle latitudes: the location of sites and regions referred to in
the text   123

8.2 The maximum extent of the last three major ice advances in Europe   124

8.3 The stratigraphy of the last 2.5×10^6 years in The Netherlands 126

8.4 The relationship between the Tenaghi-Philippon and Ioannina (Greece) pollen sequences 129

8.5 Summary of climatic reconstructions for the late glacial period in northwest Europe 132

8.6 The extent of lakes during the last ice age (Wisconsin) in the Western USA 135

8.7 The extent of loess in the USA and the loess record in central Kansas 136

8.8 Palaeotemperature reconstructions for the period 14K to 9K years BP for southern New England, Nova Scotia, New Brunswick, Newfoundland, Quebec and the south-east Baffin Shelf 137

8.9 South America, south of 30°S: location of sites and regions referred to in the text 139

8.10 South Africa, south of 30°S and the likely coastline at *c.* 18K years BP 141

8.11 Australasia, south of 30°S: location of sites and regions referred to in the text 142

8.12 The record of environmental change in Wanganui Basin, North Island, New Zealand 145

9.1 Tropical Africa: the location of sites and regions referred to in the text 149

9.2 Tropical Asia: the location of sites and regions referred to in the text 150

9.3 Changes in the configuration of the South China Sea and the vegetation zones of southern China: the situation *c.* 18K years BP, as compared with the present 153

9.4A Sea-level changes during the last *c.* 140K years, on the basis of coral terrace sequences in the Huon Peninsula, New Guinea 154

9.4B Sea-level changes during the last *c.* 390K years, on the basis of coral terrace sequences of Sumba Island, Indonesia 154

9.5 Tropical America: the location of sites and regions referred to in the text 155

9.6 The relationship between climatic change, as represented by marine oxygen isotope stages, and changes in the altitudinal position of the tree-line in the High Plain of Bogotá in the Eastern Cordillera of Colombia, as deduced from the Funza I. core 156

9.7 Tropical Australia: location of sites and regions referred to in the text 159

9.8 A summary of the results of pollen analyses from Lynch's Crater, Queensland, Australia 160

9.9 Changes in the configuration of Australia's coastline. (A) 18K years BP, (B) At present 161

Tables

2.1	The geological time scale	17
2.2	The Quaternary period and its subdivisions	25
5.1	Various types of glacial deposits	66
5.2	A summary of environmental change over the last 200K years, from Owen's Lake, California	73
5.3	The vegetation changes in East Anglia, UK, during the Hoxnian interglacial stage	74
5.4	Environmental changes of the last 12K years in Moore Lake, Alberta, Canada	76
6.1	Climatic reconstructions for Patagonia, based on tree-ring data	97
7.1	A comparison of conditions in the Iceland Sea (based on core 57-5) during the last four interglacial stages	110
7.2	Late glacial and Holocene environmental change, as indicated by marine sediments from the northern Norwegian Sea	112
7.3	The late Quaternary stratigraphy of Siberia in relation to that of Northern Europe	114
7.4	A summary of the environmental changes recorded in the sediments of Lake Lama, south-western Taymyr Peninsula, central Siberia, over the last 17K years	116
7.5	A summary of environmental change over the last 2.5×10^6 years in the Mackenzie Mountains, Northwest Territories, Canada	118
8.1	The glacial/interglacial Quaternary history of Mt Olympus, Greece	127
8.2	Generalised Holocene environmental change in the Pampas of Argentina	140
8.3	A summary of environmental change in Tasmania and western Victoria during the late Quaternary	143
9.1	Environmental change in the Nilgiris montane region, southern India, during the last $c.$ 40K years	151

Boxes

1.1 Some important theories, ideas and developments concerning Earth history and environmental change 3

1.2 The fragmentation of a supercontinent: the case of Pangaea 12

2.1 The mechanisms involved in removing carbon dioxide from the atmosphere through geological time 21

2.2 The astronomical forcing factors involved in the Milankovitch theory (astronomical theory) of climatic change 30

3.1 The principles of oxygen isotope analysis 35

3.2 The principles of alkenone stratigraphy 43

5.1 The principles of pollen analysis 70

5.2 The principles of radiocarbon age determination 81

5.3 The principles of uranium-series age-estimation methods 86

6.1 The principles involved in dendrochronology/dendroecology 92

8.1 The glaciations of the Alps 128

8.2 The late glacial period in Britain and Greenland 130–1

8.3 Changes in the prairie/forest border in the Midwest of North America during the Holocene 138

9.1 A summary of the environmental history of Mt Kenya over the last 1×10^6 years 148

9.2 A synopsis of the environmental history of the Qinghai–Xizang (Tibet)–Himalaya region 152

9.3 A summary of the palaeoenvironmental data and inferred conditions from the Huascarán ice cores, Andean Peru 158

 # Author's preface

Whilst interest in environmental change remains prevalent at all levels of education and in the popular media, there is a tendency for a concentration on human impact and the environmental issues it creates. In contrast, this book presents a survey of environmental change which has been driven by natural rather than anthropogenic factors over the last three million years.

The book comprises ten chapters. After an introductory chapter, which gives a brief history of the subject, there is a chapter on the geological history of the Earth to provide a perspective which is an essential prerequisite to any examination of environmental change during the last three million years. Chapters 3 to 6 examine the various archives of information on natural environmental change: i.e., ocean sediments, ice cores, continental archives, (for example, glacial and lacustrine sediments, peats, palaeosols, loess, carbonate deposits and packrat middens), tree rings, meteorological data and written historical records. The following three chapters examine environmental change over the last three million years on a spatial basis: high, middle and low latitudes. The conclusion emphasises the significance of patterns and periodicities in the record of natural environmental change and sets the scene for an assessment of human impact on the environment.

In accordance with the objectives of the 'Introductions to Environment' series, this text is designed with modular courses in mind, and is intended for first- and second-year students on tertiary education courses in geography, earth and environmental sciences. It provides an introduction to the causes and consequences of natural environmental change by drawing on a wide variety of methods of investigation and abundant case studies. It should also appeal to 'A'-level teachers and tertiary-level lecturers in subjects related to the earth and environmental sciences.

This field of investigation is as dynamic as environmental change itself. I hope I have captured some of the urgency and excitement associated with such studies and communicated this in a fashion that will inspire readers to pursue further studies and possibly to participate in the research itself.

A.M. Mannion
Reading, 1998

Acknowledgements

I wish to acknowledge the assistance of the many people who have contributed to the publication of this book. Susie Da Costa and Christine Holland typed the text; Judith Fox, Heather Browning and Sheila Dance drew the diagrams; Michael Turnbull prepared the final text for publication and compiled the index. This book has also benefited from the review of the manuscript by Dr Keith Crabtree, University of Bristol.

1 Introduction

1.1 Preamble

The Earth has never been characterised by a constant, static environment; environmental change has been the norm rather than a rarity throughout the *c.* 5000×10^6 years of Earth history. On both a temporal and spatial basis, the rate of change has varied, with long periods of gradual and subtle transformation punctuated by major upheavals. The study of Earth history has developed in a similar way, with periods of consolidation following the acceptance of major theories.

Environmental change has also been rhythmical in character because of the influence of regular fluctuations such as the revolution of the Moon around the Earth and the revolution of the Earth around the Sun. Such fluctuations influence the pattern of tidal frequency and seasonality. The Earth experiences additional oscillations which are cyclical, e.g. the recurrence of phenomena such as mountain building or climatic change generated by the periodicity of the Earth's orbit around the Sun (the Milankovitch hypothesis – see Section 2.4), which have occurred throughout geological time.

These and all other aspects of Earth history have influenced, and been influenced by, life. The manifestation of this relationship is the operation of **biogeochemical cycles** whereby, as is implied by the term, chemical exchanges occur between the compartments of the Earth (the lithosphere, hydrosphere, biosphere and atmosphere, Figure 1.1) and are mediated by living organisms that comprise the Earth's **biota**. Consequently, environmental change can be expressed in terms of adjustments in global biogeochemical cycles. Such changes have been ultimately associated with climatic change throughout Earth history, especially in relation to the carbon, nitrogen and sulphur cycles.

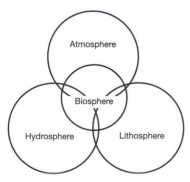

Figure 1.1 *The relationships between the lithosphere, hydrosphere, atmosphere and biosphere*

Only in the last 200 years has the idea of a dynamic Earth been in vogue. Theories concerning plate tectonics, the geographical cycle, glaciation, environmental systems and Gaia have been developed since 1850. Concomitantly, major advances have been made in field and laboratory methods which now encompass a wide range of sedimentological, biological and stratigraphic techniques. The advent of absolute dating techniques since the 1950s, when radiocarbon dating was first established, has considerably improved the precision of studies of environmental change, and has facilitated the determination of rates of change. In addition, access has been gained to hitherto inaccessible archives of information on environmental change, notably ocean sediments and polar ice. Research has also extended into remote regions such as Antarctica and high mountain regions, as well as high and low latitudes.

Interest in environmental change has intensified in the last two decades because of the focus on environmental issues such as global warming, acidification and stratospheric ozone depletion, all of which have received considerable media limelight. Such issues are no longer only the concern of scientists and academia. On the one hand they concern every member of society, and on the other hand their mitigation or control requires planning, financial assistance and expertise which involve local, regional, national and international political involvement. This reflects the close relationship between even the most technocentric members of society and the environment, as well as the interconnectedness of people, their resource use and the wider environment.

There is abundant evidence for human-induced environmental change at all scales, from the local to the global, and a growing recognition that mitigation strategies are essential to restore or preserve the status quo. Consequently, research into past environmental change, particularly that which has not been anthropogenically driven, is of considerable importance in order to obtain base-line data on natural change and rates of change. Moreover, and especially in relation to climatic change, society requires the capacity to plan for the future, i.e. to have an accurate predictive capability. One aspect of developing such a capacity is to generate models, e.g. **General Circulation Models**, whose accuracy can be evaluated to a degree by using them to simulate past natural environments. Recent natural environmental change has thus become the subject of intense investigation, especially that which has occurred in the last 3×10^6 years or so. This encompasses the Quaternary period, which is the most recent geological period coming right up to the present, and the final part of the earlier Tertiary period. The particular significance of this time interval is the repetition of periods of cold which, in the later part of the Quaternary period, were sufficiently intense to become ice ages separated by warm intervals known as interglacials. The natural environmental changes of the last 3×10^6 years are the subject of this book.

1.2 A synopsis of the development of ideas on environmental change

Ideas about environmental change have altered considerably in the last 200 years, as is reflected in the chronology given in Box 1.1. Prior to 1800, views on Earth history were dominated by the proclamation of Archbishop Ussher of Armagh in 1658. This invoked the formation of the Earth as occurring in 4004 BC, a date derived from calculations based on Biblical genealogies. A major advance came in the mid-1800s when the glacial theory was inaugurated (Emiliani, 1995). The latter was a seminal idea that represents the foundation of modern-day Quaternary studies and which had profound repercussions for the Earth Sciences in general. The recognition of the significance of cycles of erosion and uplift (the **Geographical Cycle** discussed in Section 1.2.3) and the theory of **Continental Drift** also exerted considerable influence along with developments in the biological/

Box 1.1

Some important theories, ideas and developments concerning Earth history and environmental change

		Date
1	Archbishop Ussher's statement that the Earth formed in 4004 BC.	1658
2	Catastrophism dominated ideas about Earth history.	1600s to 1850s
3	The diluvial theory, reflecting the significance of the biblical flood, was invoked to explain the occurrence of specific landforms/deposits.	1600s to 1850s
4	The glacial theory was founded by Hutton.	1795
5	Observations on contemporary glacial processes in Switzerland and Scandinavia provided evidence for the glacial theory.	early 1800s
6	Lyell proposed that icebergs were responsible for the deposition of erratics, drift, etc.	1833
7	Agassiz presented the glacial theory to the Swiss Society of Natural Sciences. This ensured the formal acceptance of the glacial theory.	1837
8	James Geikie proposed the occurrence of four glaciations in East Anglia.	1877
9	De Geer's use of varves for dating.	1880s
10	Penck and Brückner proposed four-fold glaciation for the Alpine region.	1909
11	Von Post's publication of percentage pollen analysis.	1916
12	Clements's theory of plant succession published.	1916
13	Milankovitch's astronomical theory of periodic climatic change.	1920s, 1930s
14	Concept of the ecosystem established by Tansley.	1935
15	Work began on ocean sediments.	1950s
16	Development of radiocarbon dating by Libby.	1940s/1950s
17	Adoption of the systems approach by environmental sciences.	1960s/1970s
18	Theory of plate tectonics developed.	1960s
19	Gaia hypothesis formulated.	1972
20	First deep-polar ice core (Vostok) extracted.	early 1980s

ecological sciences such as Charles Darwin's theory of evolution. Together with advances in methodology and the investigation of an increasing range of natural archives of environmental data, tremendous strides have been made in recent years. New theories such as Systems theory and the **Gaia hypothesis** have also been spawned to provide frameworks for examining the many reciprocal aspects of environmental changes, including the role of humans. There is, however, still much to learn about the complex dynamism that characterises planet Earth.

1.2.1 The glacial theory

Acceptance of Ussher's statement fostered the philosophy of **catastrophism**, whereby Earth surface features were created by cataclysmic events. Consequently, the diluvial theory was developed to explain the occurrence of many geomorphological characteristics as products of the Biblical Flood. However, by the late 1700s, the Edinburgh geologists James Hutton and John Playfair were questioning the validity of this approach. Instead, they advocated that present-day Earth surface processes could be invoked to explain the formation of landforms, and so the principle of **uniformitarianism** was established. This approach was further emphasised by Charles Lyell's promotion of the present as the key to the past. Moreover, Hutton's observations on erratic boulders in the Jura Mountains caused him to implicate glacier ice as the mechanism for transportation. It was in 1795 that Hutton laid the foundation of the glacial theory.

However, it was not until the 1820s that the glacial theory gained wider acclaim on the basis of work in the Swiss Alps by Jean-Pierre Perraudin, a mountaineer, the naturalist Jean de Charpentier, and Ignace Venetz, a highway engineer. They proposed that Swiss glaciers had once been more extensive than they were in the early 1800s. Nevertheless, there remained many sceptics who instead preferred the suggestion made by Charles Lyell in 1833 that erratics, drift deposits, etc. were deposited as icebergs melted. The formal proposal of the glacial theory came in 1837, when the Swiss zoologist Louis Agassiz presented the theory to the Swiss Society of Natural Sciences. Eventually, Lyell and other eminent natural scientists such as William Buckland, Professor of Geology at Oxford University, accepted the glacial theory. By 1860 it had achieved widespread approval and had set in train investigations that developed the theory further. For example, by 1863, Archibald Geikie, a Scottish geologist, had suggested that several glacial stages had occurred, and thus originated the idea of multiglaciation with cold periods, or ice advances, being separated by warm stages or interglacials. Archibald Geikie's brother, James Geikie, proposed in 1877 that four distinct glaciations had occurred in East Anglia, whilst evidence for multiple glaciation was found elsewhere, such as the American midWest and Europe. Of particular importance was the proposal in 1909 by Penck and

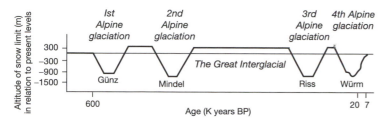

Figure 1.2 *The succession of European ice ages as determined by Penck and Brückner in 1909*

Source: Based on Imbrie and Imbrie (1979)

Brückner, two German geographers, that the river terrace sequences of the Alpine region could be ascribed to four distinct series with each terrace having been formed during one of four successive ice ages. Their scheme is illustrated in Figure 1.2.

1.2.2 Related developments

Much research was taking place in areas beyond the limits of glaciation. For example, in 1882 Ferdinand von Richthofen, a German geologist, proposed that deposits of a fine-grained yellow sediment over vast areas of Europe, Asia and North America were **loess** deposits; these originated as silt from glacial meltwaters at glacial margins and were then blown into continental interiors. The American geologist Grove Gilbert presented evidence in 1890 to show that the Great Salt Lake of Utah was a remnant of a much larger lake. This and related work gave rise to the suggestion that periods of high rainfall, i.e. pluvials, alternated with dry episodes, i.e. interpluvials, which may have been synchronous with glacial and interglacial stages respectively. Subsequently, work in low-latitude regions attempted to relate results, often erroneously, to pluvial and interpluvial stages.

It was also recognised, and formalised by the Polish geomorphologist, von Lozinski, in 1909, that as ice sheets expanded, the bordering tundra zone would also extend into areas occupied by boreal and/or temperate forest and give rise to Earth surface processes characteristic of regions close to ice sheets, i.e. periglacial processes. In addition, another Scottish geologist, Charles MacLaren, observed as early as 1841 that if large parts of the Earth experienced an expansion of ice sheets the global hydrological cycle would be altered; as water was incarcerated in ice sheets, sea-level fell, with the reverse process occurring as the ice sheets waned. Both John Playfair and Charles Lyell and later in 1865 Thomas Jamieson, a Scottish geologist, recognised the possibility that the added weight of an accumulating ice sheet could cause crustal downwarping, with a recovery on ice-sheet melting. Together, the accumulation of water as ice and the latter's weight gave rise to glacio-eustatic and glacial-isostatic changes in sea-level respectively.

Natural scientists were also turning their attention to biological evidence for environmental change. The work of Edward Forbes was amongst the earliest. In 1846 he published a British Geological Survey memoir which considered the relationship between various members of the flora and fauna, in relation to their migration into the British Isles as climate changed. Forbes also refined the term Pleistocene and inaugurated its modern usage in contrast with the earlier use of the term by Charles Lyell. Similar work by Heer in Switzerland, published in 1865, laid the foundation for the field of palaeoecology. Initially interest centred on plant macroscopic fossils (macrofossils) which are plant remains visible to the naked eye. The work of Scandinavian botanists Blytt and Sernander, in the late nineteenth and early twentieth century, involved the analysis of macrofossils from Scandinavian bogs and lakes to determine vegetation and climatic change. Notable work on British deposits, especially in East Anglia, was undertaken by Clement Reid between 1882 and 1916. Of particular importance is the development of pollen analysis in the early twentieth century. This involved, as it does today, the extraction of pollen grains and spores from peat and lake sediments and their identification to genus level. The Swedish geologist, Lennart von Post, building on earlier work by other Scandinavian geologists, such as Weber, Geinitz and Lagerheim, established the method for determining pollen counts on a quantitative basis in 1916. Since then pollen analysis has been used in palaeoenvironmental studies all over the world.

A similar approach was developed for the analysis of diatoms, unicellular algae that are the basis of the food chains in many freshwater and marine habitats. One of the earliest analyses of lake sediments was that of Crystal Lake, Wisconsin, USA, by Paul Conger in 1939, though studies of taxonomy and classification began much earlier in the 1830s. In relation to faunal remains, William Buckland produced some of the earliest work on vertebrate remains in cave earths in 1822, while Richard Owen published an early volume on British fossil mammals and birds. Molluscs, both modern and fossil, were also a source of inspiration. For example, the work of Kennard in the late 1800s and early 1900s on terrestrial and freshwater molluscs established mollusc taxonomy and ecology, essential prerequisites for his and the subsequent interpretation of molluscs in late Tertiary and Quaternary sediments. In more recent years, attention has focused on a variety of other organisms (Berglund, 1986; Lowe and Walker, 1997). Of these, research on fossil insects, and in particular Coleoptera (beetles), has proved particularly valuable. This is because some species survive only under specific environmental conditions including narrow temperature ranges. Such organisms are thus valuable palaeotemperature indicators. Although taxonomic work was being undertaken in the 1800s, it was G. Russell Coope, a geologist from Birmingham University, who began to develop the potential of beetle assemblages for temperature reconstruction in the 1960s.

As evidence on environmental change accumulated in the wake of the acceptance of the glacial theory, interest began to focus on the causes of the climatic changes that resulted in multiglaciation. The first to consider the possibility of an astronomical theory was the French mathematician Joseph Adhémar. In 1842 he proposed that changes in the orbit of the Earth around the Sun could be the underlying cause of such substantive changes in climate. Adhémar's ideas were greeted with much scepticism, but his views provided the foundation for later developments. These include the work of the Scottish geologist James Croll who suggested that ice ages occurred because the Earth revolves around the Sun in an elliptical rather than circular fashion. This had been recognised by Adhémar, but he did not consider it to be the chief driving force; this he ascribed to the precession of the equinoxes. (This is referred to in Box 2.2.) Croll's book *Climate and Time* was published in 1875, eleven years after he first formalised his hypothesis. Many eminent geologists of the time accepted Croll's ideas, but because little positive sedimentary evidence could be advanced to support them, they had been largely abandoned by 1900. However, the possibility of an astronomical cause of climatic change was revived in the 1920s and 1930s by Milutin Milankovitch, a Serbian mathematician, though it was only in the 1950s and 1960s that corroborative evidence, notably from ocean cores, was obtained. Harold Wrey, at the University of Chicago, originally discovered that the ratio of oxygen isotopes in ocean sediments is related to temperature, a characteristic that was exploited by Cesare Emiliani who identified the cycles that corresponded with Milankovitch's deliberations (Emiliani, 1955). As is discussed in Chapter 2, the Milankovitch theory is widely accepted as an explanation for climatic change, though even now there are many uncertainties about the biogeochemical mechanisms involved.

Having considered mechanisms of environmental change, natural scientists turned their attention to establishing when and at what rate such change occurred. Amongst the earliest attempts to develop precise dating, was that of the Swedish geologist Gerard de Geer in the 1880s. He recognised the deposition of annual layers of sediments in lakes close to glaciers and he developed a chronology of deglaciation based on counts of annual layers, i.e. **varves**, in the early 1900s. Varve chronologies are highly significant today. However, one of the most important advances in age determination occurred in the late 1940s, when the American physicist Willard Libby developed radiocarbon dating based on the rate of decay of radioactive carbon-14 (Box 5.2; Lowe and Walker, 1997). Many other means of age determination have been developed since the 1950s, including other radiometric methods and incremental methods such as tree-ring dating or dendrochronology (Lowe and Walker, 1997).

In relation to archives of palaeoenvironmental data, reference has been made above to glacial materials, peats, loess and lake sediments. During the 1950s, work on ocean sediments began in earnest, with the invention of corers capable of extracting long cores. The subsequent investigations of the sediments and their fossils made a major contribution to the study of environmental change, as discussed in Chapter 3. The extraction of deep ice cores from polar regions, which began in the 1960s, has made a contribution of similar importance, as is discussed in Chapter 4. The theories and methods discussed above have led to the generation of abundant information on environmental change at high, middle and low latitudes (Chapters 7, 8 and 9).

1.2.3 Additional theories of the 1800s and 1900s that influenced ideas on environmental change

Whilst the glacial theory was the source of controversy in geological circles, the publication of Charles Darwin's book *On the Origin of Species* in 1859 sparked even more controversial debate in the biological sciences. Moreover, Darwin's work influenced ideas in all branches of knowledge; not only did it revolutionise ideas about evolution, but it also influenced the earth and social sciences. In relation to the latter, debate arose as to whether or not the laws of Nature could be applied equally to society. Herbert Spencer, an English philosopher, attempted to apply Darwin's ideas on the relationship between environment and the Earth's fauna and flora to humanity. Spencer suggested that there were many similarities between organisms and societies and that to be successful only the 'fittest' in a free-enterprise system would survive. In his book published in 1864, he was the first to use the phrase 'survival of the fittest'. Spencer's ideas influenced many scholars, including geographers (Semple, 1911, in Agnew *et al.*, 1996) who began to recognise the influence of environment on the characteristics of human groups, their settlement patterns and resource use. Thus was born the doctrine of **environmental determinism**. This has been much criticised, yet elements of environmental determinism are inherent in current ideas about environmental change, as is discussed in Section 1.2.4.

However, environmental determinism greatly influenced the American geomorphologist William Morris Davis, who was Professor of Physical Geography at Harvard University from 1885 to 1912. Davis' major contribution to ideas on environmental change was his formulation of the 'cycle of erosion', which was published in 1899 and which he referred to as the 'geographical cycle'. This comprised a model of landform development that Davis believed had widespread application; he described it as a form of 'inorganic natural selection'. The model is given in Figure 1.3, and involves the continued erosion of mountains and plain formation, followed by uplift, which encourages erosion, and so the cycle is perpetuated. Davis' work has been much criticised, but it was seminal and it highlighted the dynamic nature of environmental change.

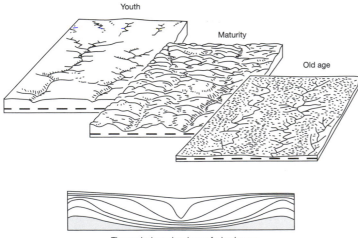

Youth

Maturity

Old age

The gradual wearing down of a landscape

Figure 1.3 *William Morris Davis' cycle of erosion*

The concept of environmental determinism is also reflected in the seminal ideas of several natural scientists concerned with plant communities (the first ecologists). Such scholars as the North American ecologists, Frederick Clements and Henry Gleason, were engaged in a polarised debate about the way in which plant communities develop. Clements, supported by Britain's Sir Arthur Tansley, considered vegetation communities to behave as 'functional organisms' involving origin, growth, the attainment of maturity and eventually death. This is similar to the youth, maturity and old age stages of Davis' 'cycle of erosion' (Figure 1.3). The deterministic element of Clements' theory was the dominant control of climate on the direction of plant community succession until it reached an optimum state that he referred to as the climatic climax community. Whilst the determinist component in this theory is explicit, so too is the dynamic disposition of vegetation communities. The theory's opponents, Moss and Gleason, also recognised the dynamism of plant communities, but advocated change based on the response of individual species rather than entire communities. The latter view is now considered correct, but both sets of views emphasise the dynamism of vegetation communities, a characteristic that has also been highlighted by palaeoecological investigations based on plant macrofossil and pollen analyses, as referred to in Section 1.2.2.

1.2.4 Recent theories on environmental change

The work of Clements and colleagues referred to in Section 1.2.3 eventually inspired Tansley to develop the concept of the **ecosystem** in 1935. This is a contraction of the phrase 'ecological system', which Tansley defined as a complex comprising both biotic components, i.e. organisms, and abiotic components, i.e. the environment in which the organisms live, including soils, atmosphere, hydrosphere and geology. The concept is thus holistic, incorporating organisms and their environment into a single framework within which individual components can be analysed. Ecosystems are structured and organised and can be identified at any scale, i.e. from the microcosm to the entire biosphere (Jones, 1997). The major components and processes that characterise ecosystems are given in Figure 1.4. This indicates the central role of organisms in fluxing materials between the lithosphere and atmosphere, as well as the nature of the biosphere as the major link between the inorganic environments that characterise the lithosphere and atmosphere. A change in any one component will promote change in all other components, which again reflects reciprocity and interdependence. This is also reflected in the Gaia hypothesis, as discussed below.

Such characteristics are typical of environmental systems in general, though it was not until the 1960s that the systems approach was adopted in the environmental sciences. The idea was developed by the Austro-Canadian biologist Ludwig von Bertalanffy in the 1930s. He linked concepts from cybernetics (the

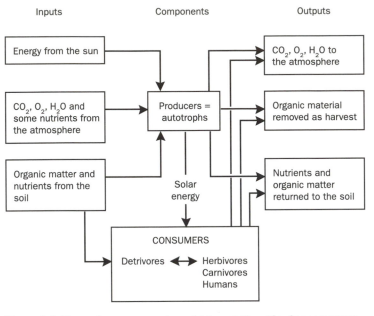

Figure 1.4 *The major components and interrelationships in ecosystems*

study of regulating and self-regulating mechanisms), which emerged as a subject within engineering in 1949, and biology, but it was only in the 1950s that his ideas began to be accepted. The approach is now considered appropriate for physical geography and environmental science, for similar reasons to those given above for the ecosystem concept. On the one hand, it recognises the complexity of the real world, i.e. holism, but on the other hand, it facilitates subdivision into identifiable components. Quantification is also possible, as well as prediction in some instances.

There are many different types of system; most environmental systems are functionally **open systems**, which means that they are characterised by the import and export of energy and matter. Natural systems are all dynamic, with exchanges and fluxes occurring continually. However, they are inherently stable and are considered to be in a state of **dynamic equilibrium**. A system is thus constant in character but active. This state of dynamic equilibrium is maintained through the operation of **feedback loops** linking inputs and outputs reciprocally. Feedbacks may be positive or negative. The operation of positive feedback causes change in the character of the system. For example, deforestation through human agency influences soil erosion and local hydrological conditions. The end result is an environmental system that is quite different to the original. Climatic change could also cause positive feedback. Conversely, negative feedback operates to maintain the character of a system; it therefore counteracts tendencies to change. In many environmental systems organisms may be involved in either positive or negative feedback.

Environmental systems can experience positive feedback over many years, possibly even millennia, so that change is subtle and imperceptible, especially in relation to a human life time. This occurs because the forcing factors are not intense and/or because systems have resilience and resistance to change; there is a lag time between receipt of a stimulus and a reaction, or because the stimulus needs to be continuous over a long period before a threshold is crossed and change becomes obvious. Natural environmental change proceeds at different rates and occurs at different scales, as does anthropogenic change. All rates and scales of change eventually alter the global environment. This is because transformations of the lithosphere, hydrosphere and biosphere – most particularly through the biogeochemical cycles of carbon, nitrogen and sulphur – have an impact on the atmosphere and thus on climate.

The relationship between atmospheric composition and the Earth's organisms is the essence of the Gaia hypothesis, which also portrays the Earth as a type of system. The hypothesis was first formulated in the early 1970s by James Lovelock, an independent scientist based in Cornwall, UK. Gaia has proved to be controversial, arousing passionate criticism and passionate praise. In common with the systems approach, Gaia focuses on the holistic nature of the Earth and particularly on the coupling of the environment and its **biota** in mutual evolution. The manifestation of this relationship is the chemistry of the atmosphere, the characteristics of which have maintained global surface temperatures in a range conducive to the support of life throughout much, if not all, of Earth history. Thus life itself has been an arbiter of environmental change, and not simply a passive receptor. Gaia also highlights the significance of biogeochemical cycles in effecting environmental change. Research on the long-term past (Section 2.2) and on recent climatic change, for example, points clearly to a vital role for the carbon cycle and its mediation by the Earth's biota. Lovelock believes that Gaia is robust and that the reciprocation between life and its environment will continue in the long-term future. However, life may not necessarily include humans, who are currently the chief cause of perturbations in biogeochemical cycles.

Mention must also be made of the theory of **plate tectonics**, which was developed in the 1960s. It capitalised on the earlier work of Alfred Wegener, a German meteorologist and geophysicist who published his theory of Continental Drift in 1915 (see discussion in Emiliani, 1995). Wegener proposed that in the early part of Earth history only one continent existed. This he named Pangaea; it later fragmented through continental drift to form the continents as they exist currently (Box 1.2). The theory of plate tectonics explains the mechanism of continental drift. This has been a major agent of change throughout Earth history, not only because of its impact on the distribution of the continents but also because of its effect on mountain building. The location of the continental masses on the Earth's surface, i.e. whether near equator or poles, as well as the presence of mountain belts and the disposition of land and sea, all have implications for global climate and the variations in climate that occur over the Earth's surface. In addition, the dispositions of land and sea, mountain and plain influence Earth-surface processes and the overall dynamics of natural environmental change. In relation to the last 3×10^6 years with which this book is concerned, the continents were in their present positions, with active mountain building in the Himalayan and Andean zones. Plate tectonics had set

Box 1.2

The fragmentation of a supercontinent: the case of Pangaea

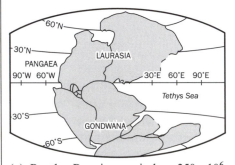

(a) By the Permian period $c.$ 250×10^6 years BP the supercontinent Pangaea was in existence. It extended from the North Pole to the South Pole.

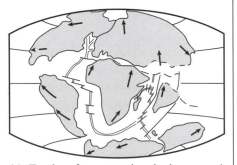

(c) Further fragmentation had occurred by the late Cretaceous period, 100×10^6 years BP. What are now South America, Africa, India and Australia had become separate landmasses.

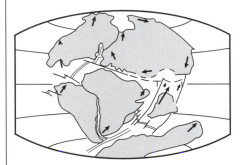

(b) Within a further 100×10^6 years BP Pangaea was fragmenting. By the late Jurassic period, 150×10^6 years ago, Laurasia had drifted north and Gondwana had drifted south.

	Midoceanic rift
	Island-arc trench
	Likely movement of lithospheric plates

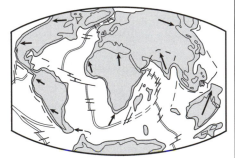

(d) Today's configuration of the continents and oceans represents additional separation of landmasses, notably the separation of Laurasia into North America and Asia. There has also been some coalescence, including the formation of the Panamanian Isthmus to join North and South America, and the northward movement of India to join Asia.

the scene on which the dramas of the great ice ages of the late Tertiary and Quaternary periods were played out. How large a role these continental configurations played in climatic change is a matter for debate, as discussed in Chapter 2.

1.3 Conclusions

The understanding of environmental change has altered considerably in the last 150 years. Indeed, it is difficult to appreciate the influence of the diluvial theory with the hindsight provided by data from terrestrial deposits, ocean sediment cores and polar ice cores. Apart from the collection of information from a wide range of archives and from widely dispersed locations, there have been major advances in understanding the links between Earth surface components and the atmosphere. The recognition that interactions between the lithosphere, atmosphere, hydrosphere and biosphere are dynamic is now firmly entrenched in all branches of environmental science. This owes its origin to the acknowledgement that observations on modern processes are the key to understanding the past. What is particularly novel about environmental research in the 1990s is the realisation that an appreciation of past environmental change is also vital to the prediction of future environmental change.

Summary Points

- Environmental change is a continuous process and has been in operation throughout Earth history.

- There are continuous interactions between the lithosphere, hydrosphere, atmosphere and biosphere.

- The replacement of the diluvial theory with the glacial theory was a seminal development in Earth/environmental sciences.

- Research on terrestrial deposits, ocean sediment cores and polar ice cores has provided the basis for a framework for global environmental change.

- Parallel developments in the biological/ecological sciences reinforced the concept of the environment as dynamic.

- Further advances include the adoption of the systems approach, which also highlights the reciprocal interactions between components of the Earth's surface.

- The Gaia hypothesis describes another type of environmental system; it emphasises the reciprocation between life and its environment in mutual evolution.

General further reading

Geosystems. An Introduction to Physical Geography. R.W. Christopherson. 1997. Prentice Hall, Upper Saddle River, New Jersey, 3rd Edition.

Ice Ages: Solving the Mystery. J. Imbrie and K.P. Imbrie. 1979. Macmillan, Basingstoke.

Physical Geography. Science and Systems of the Human Environment. A. Strahler and A. Strahler. 1997. John Wiley and Sons, New York.

Scientists on Gaia. S.H. Schneider and P.J. Boston (eds). 1991. MIT Press, Cambridge, Massachusetts.

The Ages of Gaia. J.E. Lovelock. 1995. Oxford University Press, Oxford, 2nd edn.

2 Natural environmental change: the long-term geological record

2.1 Introduction

Although this book is concerned with the last 3×10^6 years of Earth history, an adequate perspective can best be achieved if the long-term geological history is considered, albeit briefly. The geological record attests to the varied conditions that have characterised the Earth both spatially and temporally since its formation $c. 5000 \times 10^6$ years ago. Plate tectonics (Box 1.2) has caused the fragmentation and movement of continents and the creation of mountain chains, as well as the isolation and coalescence of flora and fauna to influence the course of evolution. The Earth's **biota** has altered in character as evolution has proceeded, and as continents and mountains have been moved, uplifted and eroded and new biogeographical realms have emerged. In parallel with these changes the composition of the atmosphere has altered. The reciprocal relationships depicted in Figure 1.1 were no less important in the Earth's early history than they are now.

The origin of life, the formation of the biosphere and its role in the mediation of exchanges between the lithosphere, hydrosphere and atmosphere is, however, controversial. How and when life came into existence is particularly contentious. Whilst it is clear from modern observations that life plays a vital role in the exchange of materials between the Earth's spheres (Figure 1.1), and hence the term **biogeochemical cycles**, there is much debate concerning the status of life as a forcing factor in environmental change. Traditionally, life has been considered as passive, with evolution being directed by environmental stimuli and environmental change. This premise has been challenged by the Gaia hypothesis (Section 1.2.4), which invokes a forcing role for the Earth's biota in environmental change, i.e. the biota influences other **forcing factors** such as atmospheric composition and is reciprocally influenced by this and other environmental factors. This is the approach adopted in this book.

Initially, Earth history is considered in relation to the long-term changes in atmospheric composition that have occurred, and how such changes have influenced and been influenced by life as well as major plate movements. Thereafter, a brief description of environmental change during the early and later periods of geological history is given, with emphasis on periods of ice age occurrence. Such considerations set the scene for environmental change over the last 3×10^6 years, the framework for which is discussed in the final section of this chapter.

2.2 Long-term geological history

The chemistry, physics and biology of the Earth's rocks provide the wherewithal for reconstructing its geological history. The techniques that facilitate environmental reconstruction are varied and often highly sophisticated, ranging from field observations to automated chemical analyses and **radiometric age determination**. The latter, which involves the determination of decay particles from radioactive elements, has enabled the establishment of ages for the beginning and end of the aeons, eras, periods and epochs that comprise the geological time scale. This is given in Table 2.1. The oldest rocks of the lithosphere which have so far been discovered are ancient **gneisses** from north-western Canada. These are dated at 3960×10^6 years. However, these rocks do not reflect the age of the Earth, because in its initial state it is likely that there was no solid crust; rocks formed soon after the Earth's formation may also have dissolved again into the molten interior, leaving no record. Indeed it is generally accepted that the Earth was a molten mass at its point of origin. This is estimated to have occurred *c.* 4600×10^6 years ago: a date obtained from various types of meteorites with the same chemical composition as the Earth, which thus suggests a common origin.

The first 1000×10^6 years of Earth history was probably turbulent. Temperatures were in the order of 800°C, and there was much bombardment by meteorites, though the evidence for this was destroyed as the molten surface altered and cooled. Thereafter, the crust formed (Figure 2.1) from the relatively lightweight elements that remained after the separation of heavy elements which sank to the Earth's centre, the core. On cooling, the lightweight elements formed the crust, which rests on the mantle. This has a plastic constituency and contains convectional cells. Movements of these cells cause the crust to move and crack. The large units are the plates (Box 1.2), which, throughout geological history, have fused to form so-called supercontinents, or fragmented to form a number of

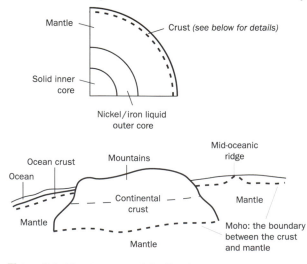

Figure 2.1 *The structure of the Earth*

Table 2.1 *The geological time scale*

Aeon	Era	Period	Epoch	Approx date of start × 10^6
	Cainozoic	Quaternary	Holocene	
			Pleistocene	1.8
		Tertiary	Pliocene	
P			Miocene	
H			Oligocene	36.6
A			Eocene	
N			Palaeocene	
E				
R	Mesozoic	Cretaceous		135
O		Jurassic		
Z		Triassic		
O				
I	Palaeozoic	Permian		290
C		Carboniferous		
		Devonian		
		Silurian		
		Ordovician		
		Cambrian		
P	Precambrian			570
R				
O				
T				
E				
R				
O				
Z				
O				
I				
C				2500
A				
R				
C				
H				
A				
E				
A				
N				4600

continents, such as those that characterise the Earth today. Plate tectonics is a major cause of long-term environmental change and contributes to short-term change through the occurrence of earthquakes and volcanic activity. Both of these activities occur along plate boundaries, which are also the site of mountain building if two continental plates converge.

Apart from plate tectonics, two additional factors have contributed substantially to long-term environmental change. One of these is life. When it occurred and how it occurred remain unresolved issues. Life dates back at least 3500×10^6 years. According to Davis and McKay (1996), there are three groups of theories to explain life's origins. One group focuses on the coalescence of organic molecules on Earth itself, another proposes formation from inorganic materials, whilst the third invokes an extraterrestrial origin. In all cases water is essential and, unlike many other planets, Earth has an abundance of water, the oceanic volume of which has been similar to that of the present for 3500×10^6 years. The second factor that has a reciprocal relationship with long-term environmental change is the composition of the atmosphere. In the early years the Earth had an atmosphere dominated by water vapour, carbon dioxide, nitrogen and possibly sulphurous gases. This is a very different composition to that of today. Over time the proportions of nitrogen, and particularly of oxygen, have increased, whilst the amounts of carbon dioxide and sulphurous gases have declined. This reflects the operation of **biogeochemical cycles** and a long-term relationship between the lithosphere and the atmosphere. It also reflects the role of organisms, and hence the biosphere, in biogeochemical cycles, and the reciprocity that occurs between the lithosphere, hydrosphere, atmosphere and biosphere (Figure 1.1).

2.2.1 Major changes in Earth history: atmospheric composition, life and temperature

Atmospheric composition has changed considerably throughout geological history. This is particularly important because of its role in determining global temperatures; in turn this influences the distribution of life forms on the Earth's surface and the very persistence of life since it evolved c. 3500×10^6 years ago (see review in Mannion, 1997a). While all gases are important in terms of surface temperatures, as is evidenced by today's concerns about the roles of nitrous oxide, methane, water vapour, etc., as heat-trapping gases, the changing concentrations of carbon dioxide and oxygen are considered here as key factors in long-term environmental change.

Figure 2.2 summarises the major environmental changes that have occurred throughout Earth history. The relationships are complex, with the evolution of life playing a key role in determining atmospheric composition. In a reciprocal relationship, i.e. **Gaia** (Section 1.2.4), atmospheric composition has been a major

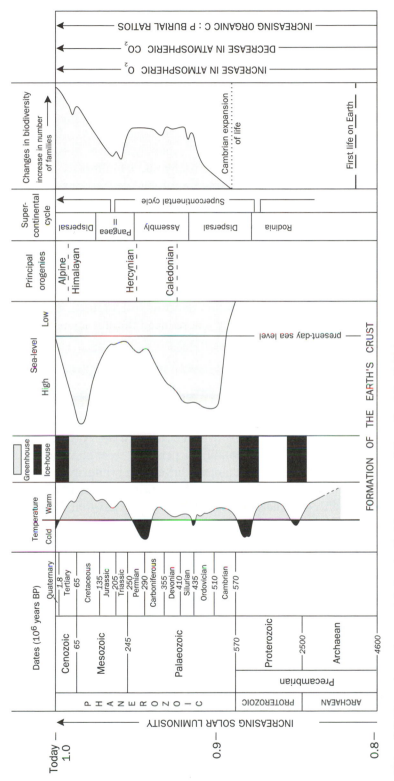

Figure 2.2 *The major environmental changes that have occurred throughout Earth history*

Source: Based on sources quoted in the text.

influence on the evolution of life. With the advent of life and its development, temperatures at the Earth's surface have been maintained within a limited range (c. 10°C to 20°C with an average of 15°C). This, in turn, is favourable to the maintenance of life. Consequently, the biosphere has been a major component and regulator of the Earth's physiology, i.e. geophysiology as defined in Lovelock's Gaia hypothesis (Lovelock, 1995).

This equilibrium in temperature has been achieved through several mechanisms which involve global biogeochemical cycles. Moreover, this equilibrium has been attained despite a gradual increase in the Sun's luminosity, which has a warming effect. Amongst these mechanisms are the silicate–carbonate buffering system and the activity of photosynthesising organisms. As Box 2.1 shows, both of these mechanisms conspire to reduce the amount of carbon dioxide in the atmosphere, incarcerating carbon compounds so produced in carbonate-rich sediments that eventually became limestone rocks, and in organic matter. Additional mechanisms operate to bury some of the organic matter and so remove it from active circulation. These involve the formation of carbon–phosphate complexes and their interment in sedimentary deposits (see Figure 2.2). This removal of carbon dioxide from the atmosphere reduced the efficacy of the greenhouse effect, and so has prevented the Earth from heating up.

Changes in the concentration of oxygen in the atmosphere have also played a significant role in environmental change. Although the silicate–carbonate system referred to above (Box 2.1) consumes oxygen, the amount of oxygen in the atmosphere has increased by a factor of a thousand since the Archaean (Figure 2.2), the atmosphere of which contained no oxygen. Concentrations similar to those at present, notably 21 per cent, were attained 250×10^6 years ago (Berner, 1994). Two mechanisms have been primarily responsible for this: the burial of organic carbon and the evolution and spread of organisms capable of photosynthesis. The equations in Box 2.1 that describe these processes show that oxygen is released. The earliest organisms capable of photosynthesis were types of blue-green algae (Cyanophyta); they were capable of surviving in an oxygen-poor world but produced oxygen through photosynthesis. They appeared c. 3500×10^6 years ago and include **stromatolites**.

As Figure 2.2 shows, the Cambrian period witnessed a major expansion of organisms, including single-celled and multicellular groups. The expansion continued into the Palaeozoic era, with a major expansion of vascular plants occurring during the Devonian period c. 410×10^6 years ago. Berner (1994) suggests that as the major colonisation of the continents by vascular plants ensued, giving rise to vast coal-forming swamps, (i.e. the coal, oil and natural gas used today as fossil fuels), the concentration of carbon dioxide in the atmosphere declined from c. 4500 ppm to c. 370 ppm. This probably contributed to global cooling, which occurred toward the end of the Carboniferous period

---

Box 2.1

The mechanisms involved in removing carbon dioxide from the atmosphere through geological time

A The silicate–carbonate buffering system

$$CaSiO_3 \;+\; CO_2 \;\underset{\text{weathering}}{\overset{\text{chemical}}{\rightleftharpoons}}\; CaCO_3 \;+\; SiO_2$$

calcium   carbon   calcium   quartz
silicate   dioxide   carbonate

$$MgSiO_3 \;+\; CO_2 \;\underset{\text{weathering}}{\overset{\text{chemical}}{\rightleftharpoons}}\; MgCO_3 \;+\; SiO_2$$

magnesium   carbon   magnesium   quartz
silicate   dioxide   carbonate

These chemical processes cause the removal of *c.* 80 per cent of the carbon dioxide produced as a result of volcanic eruptions. Throughout geological time, these mechanisms have contributed to reducing the amount of carbon dioxide in the atmosphere and hence to the diminution of its greenhouse effect.

B Photosynthesis

$$6CO_2 \;+\; 6H_2O \;\xrightarrow{\text{light energy}}\; C_6H_{12}O_6 \;+\; 6O_2$$

carbon   water   carbohydrate   oxygen
dioxide

This process, which is the domain of chlorophyll-bearing plants, also contributes to the removal of carbon dioxide from the atmosphere. The resulting organic compounds provide an energy source for the green plants themselves and for all heterotrophic organisms. Equally important is the production of oxygen, which enters the atmosphere. Without oxygen in the atmosphere, mammals would not have evolved.

---

when the Permo-Carboniferous glaciation developed (Section 2.2.2). The development of this and other ice ages, as well as those of the Quaternary period (Section 2.3), reflects the fact that Earth's surface temperatures have not been constant. Many other factors have contributed to temperature change, as discussed below, but despite periods of biotic extinctions, life has continued to proliferate.

2.2.2 Supercontinents, ice houses and greenhouses

The operation of plate tectonics (Box 1.2; Section 2.2) throughout geological history has caused continents to drift apart and to coalesce repeatedly. Professor J. Tuzo Wilson of the University of Toronto recognised that tectonic events generated cycles, which have come to be known as Wilson cycles or supercontinental cycles. There have been several such cycles during the Proterozoic and Phanerozoic aeons (Figure 2.2). Two such supercontinents include Pangaea (see Box 1.2) which had formed by $c. 200 \times 10^6$ years ago, and an earlier supercontinent known as Rodinia, which formed $c. 700 \times 10^6$ years ago. It is also likely that at least one, and possibly several, supercontinents existed during the Archaean.

The disintegration of Pangaea is illustrated in Box 1.2, which shows an initial split into a northern continent known as Laurasia and a southern continent known as Gondwana. This occurred $c. 250 \times 10^6$ years ago. By the late Cretaceous, $c. 100 \times 10^6$ years ago, further subdivisions had occurred and the configuration of the continents became similar to that of today, though it was not until $c. 50 \times 10^6$ years ago that North America and Eurasia drifted apart. The breakup of Rodinia followed a similar pattern, as the supercontinents of the Archaean may have done.

These evolutionary changes have involved periods of **orogenesis**, i.e. mountain building as plates converge, and periods of increased submergence as coastlines have been created through continental disintegration. The state of emergence or submergence influences the rate of organic carbon burial (Section 2.2.1), and thus has an impact on climatic change. For example, during the emergent stages of supercontinents, when sea-levels are falling, erosion and weathering release calcium, magnesium and phosphorus. As a result carbon sequestration increases, a diminished greenhouse effect ensues and the concentration of oxygen in the atmosphere increases. These changes give rise to what are referred to by Worsley *et al.* (1991) as 'biotic innovations'. These are bursts of evolution, caused possibly by the creation of an enhanced range of ecological niches. This reinforces carbon dioxide removal. If such trends continue, and if the Earth is at its most distant from the Sun (see discussion on the Milankovitch theory of climatic change, Section 2.4), an ice age may develop because of the reduced greenhouse effect as carbon dioxide is removed from the atmosphere.

Moreover, there have been periods of mass extinction. These are shown in Figure 2.3. It is possible that they have occurred at regularly spaced intervals, though no common cause has been pinpointed and the idea has received much criticism. In terms of evolutionary history, mass extinction events are often followed by phases of biotic expansion; the niches vacated as extinction occurred, and the new niches created through environmental change, both provide opportunities for remaining organisms to adapt, evolve and expand their populations. The periods of mass extinction may correspond with periods of supercontinent formation. At such a

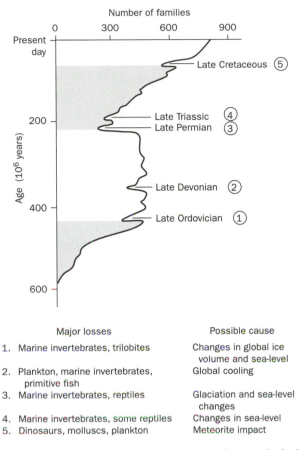

Number of families

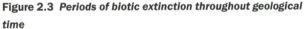

Major losses

1. Marine invertebrates, trilobites
2. Plankton, marine invertebrates, primitive fish
3. Marine invertebrates, reptiles
4. Marine invertebrates, some reptiles
5. Dinosaurs, molluscs, plankton

Possible cause

Changes in global ice volume and sea-level
Global cooling

Glaciation and sea-level changes
Changes in sea-level
Meteorite impact

Figure 2.3 *Periods of biotic extinction throughout geological time*

stage the shallow-water marine habitat would diminish in extent. This would be exaggerated if a marine regression occurred. The loss of habitats alone would cause considerable extinction, but if buried organic matter were to oxidise through weathering and erosion, a global climatic event might be precipitated. The release of carbon would reduce the amount of free oxygen which alone could cause widespread extinction in the oceans and on the continents. (Note that this possibility contradicts the hypothesis of Worsley *et al.* (1991) referred to above.) This change in atmospheric composition could cause global warming. Other suggestions for mass extinction include global cooling and the impact of large meteorites. This latter idea has gained acceptance as the reason for the demise of the dinosaurs at the Cretaceous–Tertiary boundary.

Figure 2.2 shows that there have been at least three periods when the Earth experienced 'ice house' conditions during the Phanerozoic aeon and possibly two during the earlier Proterozoic. The earliest dates back to 2250×10^6 years, with a second occurring *c.* 850×10^6 years. Both were periods of extensive glaciation. According to recent research by Evans *et al.* (1997), there is convincing evidence for extensive glaciation in tropical as well as polar regions. Between 570×10^6 and 450×10^6 years ago, global warming ensued to produce greenhouse conditions; at the same time there was a major expansion of life forms. The Earth was cooling again by 450×10^6 years ago, and ice-house conditions prevailed for *c.* 30×10^6 years; average global temperatures declined to *c.* 22°C from *c.* 40°C and extensive ice sheets developed in the polar regions.

For the next $100 \times \times 10^6$ years, a warm but dry environment prevailed, and this was followed by the Permo-Carboniferous ice age. This lasted for *c.* 70×10^6 years (330×10^6 to 260×10^6 years ago). Sea-levels were especially low and it is likely that glaciers and ice sheets expanded and regressed cyclically, just as they did during the late Tertiary/Quaternary ice ages. A greenhouse period of *c.* 200×10^6 years separated these two periods of ice advance and retreat. This 200×10^6 years was far from uniform in terms of global climate; a cool period prevailed between 187×10^6 and 105×10^6 years, with warm periods before and after. In the Cretaceous period greenhouse conditions were at their optimum and a period of flowering plant (angiosperm) diversification occurred (Crane *et al.*, 1995). The actual onset of the most recent ice-house period is a matter for debate. Although there is evidence for actual glaciation only *c.* 3.5×10^6 years ago, it is generally considered that cooling actually began 55×10^6 years ago and gradually intensified. Additional detail is given below in Section 2.3.

2.3 The last three million years: the short-term geological record

The geological time scale given in Table 2.1 indicates that the last 3×10^6 years of Earth history comprises the later part of the Tertiary period (the later part of the Pliocene epoch) and the Quaternary period. This assumes that the base of the Quaternary is 1.8×10^6 years old, a proposal promoted by Berggren *et al.* (1995) but which is not accepted throughout the geological community. As discussed in Section 1.2.1 in relation to the development of the glacial theory, the term 'Quaternary' came to become synonymous with the ice ages. A date of 1.8×10^6 years post-dates unequivocal evidence for the onset of glaciation which, on this basis, began during the Pliocene epoch. Certainly, there is evidence for glaciation as long ago as *c.* 38×10^6 years in Antarctica (Webb and Harwood, 1991) which reflects the establishment of a cooling trend that began in the Tertiary period. Moreover, there is widespread evidence for the onset of glaciation in the North Atlantic region at *c.* 3.5×10^6 years and 2.45×10^6 years (reviewed in Mannion, 1997b). The latter corresponds with changes in the fossil record, i.e. biostratigraphic changes, in The Netherlands, reflecting climatic deterioration, and in ocean sediment cores from both the North Atlantic and North Pacific (Morley and Dworetzky, 1991; Versteegh, 1997). Terrestrial loess sequences in China and several deep lake sediments from various parts of the world also provide evidence for climatic cooling at around this time. Consequently, many researchers believe that the base of the Quaternary should be placed at *c.* 2.4×10^6 to 2.6×10^6 years.

The official view, that of the International Commission on Stratigraphy (Cowie and Bassett, 1989), has formally adopted a stratigraphic horizon in a geological section at Vrica in Calabria, Italy, as the base of the Quaternary. Here, a claystone overlies a sapropel bed (a black mud rich in organic calcium carbonate). This junction is immediately below the initial appearance of *Cytherpteron testudo*, a cold-loving ostracod. This horizon is 1.8×10^6 years old. Changes of a similar age have been found in many ocean cores and continental deposits, though the cooling reflected in these deposits is not always as intense as that which occurred *c.* 2.5×10^6 years ago. In addition, there is a dispute over the identification of the ostracod, which may not, therefore, reflect cooling (Jenkins, 1987). The major geological subdivisions of the last 3×10^6 years are given in Table 2.2, which is based on an age of 1.8×10^6 years for the base of the Quaternary.

Table 2.2   The Quaternary period and its subdivisions

Approximate age (K years BP)	Period (or system)	Epoch (or series)	Other subdivisions	Palaeomagnetic polarity epochs	Events
0					
10	Q	Holocene			
	U	P			
	A	L	Upper		
	T	E	Pleistocene		
125	E	I			
	R	S	Middle		
	N	T	Pleistocene		
	A	O			
750	R	C		Brunhes	
	Y	E	Lower	———	Jaramillo
		N	Pleistocene	Matuyama	
		E			Olduvai
1800	↑	↑			
2010	TERTIARY	PLIOCENE		Gauss	Réunion
5000	↓	↓			

Since the inception of the glacial theory (Section 1.2.1), and the recognition of glacial deposits, attempts have been made to correlate between regions and continents. However, the task is difficult, not least because it is only recently that new age-determination techniques have been developed to extend the range of 50×10^3 years, which is all that is possible with radiocarbon age estimation. Moreover, correlation over even short distances is problematic, because of the fragmentary nature of the continental geological record and the spatial juxtaposition of glacial, periglacial and temperate deposits.

Currently, the emphasis is on the establishment of regional sequences, each with its own terminology. Thereafter, attempts have been made to relate individual sequences to those elsewhere, using a variety of criteria, in order to begin to develop a framework of global environmental change. Such an approach is illustrated in Figure 2.4, which gives correlations for much of the Northern Hemisphere. The advent of ocean-core stratigraphy and dating has led to the establishment of a series of marine stages (Chapter 3), to which terrestrial, lacustrine, ice and loessic records (Chapters 2, 3 and 4) can be related.

In the Southern Hemisphere, there are far fewer terrestrial deposits from the last 3×10^6 years than in the Northern Hemisphere, and research has not been under way for so long. Nevertheless, ocean-sediment cores and Antarctic ice cores, as well as lacustrine and terrestrial deposits, are providing important information on environmental change. Similarly, the humid and arid tropics are beginning to yield their palaeoenvironmental secrets. Overall, the records reflect a highly dynamic Earth, which switched from warm to cold stages with considerable regularity. During the last 3×10^6 years there were *c.* 50 or more cold or **glacial stages** and

	NORTH EUROPEAN STAGES	UK	NETHERLANDS	NORTH GERMANY	POLAND	USSR	ALPINE REGION (AUSTRIA & GERMANY)	NORTH AMERICA	MARINE STAGES	DATES (K YEARS BP)
IG	HOLOCENE	FLANDRIAN	HOLOCENE	HOLOCENE	HOLOCENE	HOLOCENE	HOLOCENE	HOLOCENE	1	13
G1	WEICHSEL	LATE DEVENSIAN / MIDDLE DEVENSIAN / EARLY DEVENSIAN	WEICHSELIAN	WEICHSEL GLACIATION / LOW TERRACE	MAIN STADIAL / PREGRUNDZIAD STADIAL / KASZUWSTADIAL / KASZULY STADIAL (VISTULIAN)	LATE VALDAI G. / LOESS FLUVIAL DEPOSITS	MAX WURM G. STILLFIED B. / FIRST WURM G. / STILLFIED A.	LATE WISCONSIN / MIDDLE WISCONSIN / EARLY WISCONSIN / EOWISCONSIN	2, 3, 4, 5 (a, b, c, d)	32-35, 64-65, 75-79
IG1	EEM	IPSWICHIAN	EEMIAN	EEMIAN	EEMIAN	MIKULINO	MONDSEE & SOMBERG	SANGAMON	5 e	122, 128-132
G2	WARTHE	RIDGEACRE? (WOLSTONIAN COMPLEX)	(SAALIAN COMPLEX)	SAALE 3 G. / RUGEN 1.G. / SAALE 2 G.	WARTA GLACIATION	MOSCOW (DNEIPER STAGE)	LATE RISS / LATE RISS / RISS? / LATE RISS TEERACE	LATE ILLINOIAN	6	195-198
1G2	SAALE / DRENTHE	STANTON HARCOURT	HOOGEVEN	TREENE / WENNINGSTEDTER & KITTMITZER PALAEOSOLS	LUBLIN I.G. / POLICHNA I.S.	ODINTSOVU I.S.	PARABRAUN-EARTH	ILLINOIAN	7	251-262
G3	DRENTHE		DRENTHE GLACIATION	DRENTHE G. / MAIN TERRACE	ODRA G. / PODWINEK I.S. / PRE-MAXIMUM STADIAL (CENTRAL POLISH GLACIATION)	DNEIPR GLACIATION	ANTEPENULTIMATE G. / EARLY RISS / MINDEL / HIGH TERRACE	EARLY ILLINOIAN	8	297-302
1G3	DOMNITZ (WACKEN)	HOXNIAN		FREYBURGER BODEN	MAZOVIAN	ROMNY	REDDISH PARABRAUN-EARTH		9	
G4	FUHNE (MEHLECK)			ELDERITZER TERRACE / ERKNER ORGANIC SEDIMENTS	PRE-MAXIMUM STADIAL? / WILGA G.?	ORCHIK STAGE (PRONYA GLACIATION)	GL4 TERRACE (PRE-RISS TERRACE)		10	338-347
1G4	HOLSTEINIAN (MULDSBERG)	SWANSCOMBE	HOLSTEIN	HOLSTEIN	FERDYNANDOW	LICHVIN	REDDISH PARABRAUN-EARTH		11	367-352, 440-428

Code					South Polish Glaciation	Oka Glaciation	Alps	MIS	Age
G5	ELSTER 2	ANGLIAN	ELSTER	ELSTER 2 G.	SAN G.	OKA GLACIATION	GL5 GLACIATION OF ALPS / LATE MINDEL / DONAU MINDEL GL5 TERRACE	12	472–480
1G5	ELSTER 1/2	CROMERIAN	ELSTER	FLUVIAL GRAVELS VOIGSTEDT	KOCK STADIAL?	I.G.?	RIESENBADEN PALAEOSOL	13	502–?
G6	ELSTER 1			ELSTER 1 G.	LUSZANA I.G.? NALECZORIG SERINKI STADIAL?		LOWER INTERTERRACE GRAVEL / EARLY MINDEL / DONAU	14	542–562
1G6	CROMERIAN IV	WAVERLEY WOOD	CROMERIAN IV	FLUVIAL GRAVELS VOIGSTEDT	PODLASIE 1.G.		RIESENBODEN PALAEOSOL	15	592–630
G7	GLACIAL C	K E S G R A V E	GLACIAL C	FLUVIAL GRAVELS	PRE-CROMERIAN GLACIATIONS		MIDDLE INTERTERRACE GRAVEL	16	627–687
1G7	CROMERIAN III		CROMERIAN III				RIESENBODEN PALAEOSOL	17	647–718
G8	GLACIAL B		GLACIAL B				UPPER TERRACE GRAVEL GL8 GLACIATION	18	790
1G8	CROMERIAN II	G R O U P ?	CROMERIAN II (WESTERHOVEN)				RIESENBODEN PALAEOSOL	19	
G9	HELME (GLACIAL A)		GLACIAL A	HELME FLUVIAL GRAVELS			EARLY GUNZ?	20	
1G9	ASTERN INTERGLACIAL		WAARDENBURG	UPPER MUSHELTONE			INTERGLACIAL	21	

PRE-ILLINOIAN

Figure 2.4 *Correlations between various Quaternary sequences in the Northern Hemisphere*

Source: Based on Mannion (1997b) and Lowe and Walker (1997).

another 50 warm or **interglacial** stages. The latter were periods of relatively short duration, *c.* 20×10^3 years, during which temperatures were similar to those of the present day or a few degrees Celsius higher. The cold or glacial stages were relatively long lived; before *c.* 900K years BP each glacial stage lasted *c.* 40K years and after *c.* 900K years BP they lengthened to *c.* 100K years. Temperatures were then as much as 10°C lower than they are today. Not all such episodes were characterised by the accumulation and advance of polar and high-altitude ice sheets; some were much colder than the present day but not cold enough for extensive ice-sheet formation.

These cold and warm periods were not internally uniform in temperature. There is abundant evidence for considerable variation within each episode. In glacial stages, for example, there were often relatively short-lived warm periods, which are known as **interstadials** i.e. warm periods occurring within glacials (between ice advances or **stadials**). Moreover, the oscillation between warm and cold stages prompted major alterations in the configuration and content of the world's ecosystems. The Earth's biosphere was quite different during interglacials and glacials. The atmosphere and hydrosphere were also quite different. The interglacial atmosphere was carbon-dioxide and methane rich, whilst that of glacials was impoverished. This links the biosphere and atmosphere through the carbon cycle.

In relation to the hydrosphere, glacial stages witnessed the incarceration of huge volumes of the Earth's water resources in ice caps. Consequently, such stages were characterised by large falls in sea-levels, possibly by as much as 150 m. Continental shelves were exposed and coral reefs ceased to grow; arid conditions prevailed in the continental interiors. The rapidity, geologically speaking, with which such oscillations have occurred has attracted attention for some time (Section 1.2), but has become especially pertinent in the last few decades as anthropogenic global warming has become increasingly likely. An examination of past changes and rates of change, plus the determination of biosphere–hydrosphere–lithosphere–atmosphere reciprocities, is vital to understanding and predicting future environmental change.

2.4 The causes of climatic change

The environmental changes outlined above were underpinned by climatic change. As discussed in Section 2.2, climatic change has been a characteristic of planet Earth since it first came into existence, and many theories have been proposed to explain it. Until recently, most such theories considered climatic change as a separate entity that forced change in the lithosphere, biosphere, etc. Now, however, it is acknowledged that climatic change can be generated from within the Earth system (see Figure 1.1), because of the reciprocity that exists between the various components. This is well expressed in Lovelock's **Gaia** hypothesis, which was discussed in Section 1.2.4. Nevertheless, there is evidence that external factors are also important. These include the way in which the Earth rotates around the Sun, and also the impacts of meteorites. Some of the theories concerning climatic change have been referred to in Section 2.2, e.g. mountain building and biotic radiations. It is likely that most climatic change is due to more than one cause, because of the complex interrelationships between the Earth's components. Moreover, it is difficult, if not impossible, to determine which factors are **forcing factors**, i.e. which cause positive feedback (Section 1.2.4), and which are **reinforcing factors**, i.e. factors that contribute to change when they themselves have been affected by initial change.

Amongst the most accepted causes of climatic change is the Milankovitch theory (Section 1.2.2). Whilst this theory was formulated in the 1930s, it was reinstated as a serious proposition by the work of Hays *et al.* (1976), which recognised Milankovitch's three main cycles in the oxygen isotope records of ocean-sediment cores (Chapter 3). The theory of Milankovitch is explained in Box 2.2. In particular it is considered that changes in orbital eccentricity drive the glacial–interglacial cycle. The cycles of axial tilt influence the pattern of stadials; the cycle involving the precession of the equinoxes may cause interstadials to develop (reviewed in Imbrie *et al.*, 1993). Thus the way in which the Earth revolves around the Sun influences climatic change, because it causes the amount of solar radiation received at the Earth's surface (insolation) to vary. This is particularly important in the high latitudes of the Northern Hemisphere. However, orbital eccentricity has only a limited effect on insolation and it is now considered that other factors must contribute to global cooling. Possibilities include changes in oceanic circulation, caused by a variety of mechanisms, and changes in the composition of the atmosphere: notably the depletion of greenhouse gases. Certainly, the rise and fall in the concentrations of carbon dioxide and methane that occur in polar ice cores (Chapter 4) parallel episodes of global cooling and warming. Possibly Milankovitch's orbital forcing creates positive feedback, which causes the biosphere, or the oceans, or both, to act in a reinforcing manner by absorbing greenhouse gases.

Other possibilities for contributing to climatic change are tectonic activity/land uplift and oscillations in sun-spot activity. The latter exhibit 18.6 year and 11 year cycles, reflecting oscillations in the amount of heat energy emitted by the Sun (reviewed in Mannion, 1997b). It is, however, unlikely that such oscillations would precipitate major climatic change; indeed, in view of the many claims that have been made for correlations between sun-spot cycles and floods, plagues, droughts, famines, etc., it is difficult to take sun-spot cycles seriously. On the other hand, **orogenesis** may well contribute to climatic change. In particular it is important to consider why the ice ages of the last 3×10^6 years developed at all, and is there a cause shared by these and earlier ice ages? Raymo and Ruddiman (1992) and Raymo (1994) have suggested that the ice ages were prompted by tectonic uplift which resulted in enhanced weathering, i.e. the breakup of rocks

Box 2.2

The astronomical forcing factors involved in the Milankovitch theory (astronomical theory) of climatic change

A Orbital eccentricity

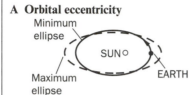

Minimum ellipse

Maximum ellipse

Periodicity *c.* 96K years

The Earth's orbit around the Sun varies, and is elliptical rather than circular. When the Earth is furthest from the Sun, cooling occurs. The periodicity of orbital eccentricity is considered to be a major factor in the waxing and waning of ice ages.

B Axial tilt

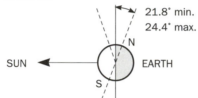

21.8° min.
24.4° max.

SUN   EARTH

Periodicity *c.* 42K years

The tilt of the axis around which the Earth rotates causes its seasonality. It also determines the intensity of incident radiation. When the angle of tilt is at its minimum, 21.8°, incident radiation in the Northern Hemisphere is *c.* 15 per cent less than when the angle of tilt is at its maximum, 24.4°. Periods of minimum tilt therefore relate to cooling.

C Precession of the equinoxes

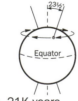

Equator

Periodicity *c.* 21K years

This occurs due to the wobble of the Earth's axis. It controls the amount of solar radiation received at the Earth's surface by influencing the season in which the Northern Hemisphere is closest to the Sun. In particular, an ice age is likely to develop when the Northern Hemisphere is furthest from the Sun in summer.

D Variations in solar radiation resulting from A, B and C above

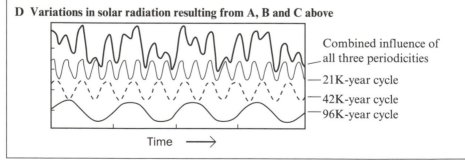

Combined influence of all three periodicities

21K-year cycle
42K-year cycle
96K-year cycle

Time ⟶

through physical and chemical means. They suggest that the uplift of the Tibetan and Colorado Plateaux in the Pliocene epoch could have altered the structure of the air streams of the upper atmosphere, i.e. the **jet streams**, resulting in the shift of cold air from high to mid-northern latitudes. This cooling of landmasses coincided with additional cooling caused by orbital forcing. Raymo and Ruddiman (1992) also indicate that increased chemical weathering of silicates, enhanced by uplift, consumed atmospheric carbon dioxide and so reduced the greenhouse effect. Add to this the effect of the closure at $c.\ 3 \times 10^6$ years ago of the Panamanian Isthmus to separate the Atlantic and Pacific oceans, and it is clear that between 3.5×10^6 and 2.5×10^6 years numerous changes were occurring that could have all contributed to the onset of the Tertiary/Quaternary ice ages.

2.5 Conclusion

Current understanding of the Earth's geological history, both long and short term, highlights its dynamic history during the last 5000×10^6 years. There is still much to learn. The long-term past is more difficult to unravel than the recent past, because the evidence is destroyed by more recent environmental change. Particular enigmas concern the mechanisms for the origins of the Earth itself and of life.

The theory of plate tectonics provides a sound basis to explain long-term change in general, though not in detail. Why the Earth should have alternated between an ice-house and greenhouse state during the last 570×10^6 years (and probably for part of the Proterozoic aeon as well) has not been established with certainty. How periods of extinction and evolutionary radiation relate to these oscillations and to supercontinent coalescence and fragmentation also remains to be determined.

Just as many questions pertain to how and why environmental change has occurred during the last 3×10^6 years. The reasons for the waxing and waning of ice ages are both enigmatic and fascinating. The operation of Milankovitch cycles is now acknowledged, but this is only part of the story. Some of the detail will be explored in the ensuing chapters.

Summary Points

- Environmental change has characterised the Earth's 5000×10^6 year history.
- There is little evidence for environmental change during the first 2000×10^6 years when the Earth was still in a molten state.

- Plate tectonics provide a model for environmental change, involving the fragmentation and coalescence of supercontinents such as Pangaea.

- The Phanerozoic aeon was characterised by a series of alternating ice-house and greenhouse conditions.

- Life and the biosphere are intimately related to environmental change throughout geological history (**Gaia**).

- The last 3×10^6 years have witnessed the onset of **glacial**/cold stages and **interglacial**/warm stages.

- Milankovitch's mechanisms (the astronomical theory) of climatic change are considered to be the prime causes of ice ages.

- Mountain building, changes in oceanic circulation and changes in atmospheric composition are likely contributors to climatic change.

General further reading

Global Environmental Change. Its Nature and Impact. J.J. Hidore. 1996. Prentice Hall, Upper Saddle River, New Jersey.

Global Environmental Change. Past, Present and Future. K.K. Turekian. 1996. Prentice Hall, Upper Saddle River, New Jersey.

Life as a Geological Force. P. Westbroek. 1991. W.W. Norton and Co., New York.

Plate Tectonics. A.N. Strahler. 1997. John Wiley and Sons, New York.

The Key to Earth History. P. Doyle, M.R. Bennet and A.N. Baxter. 1995. John Wiley and Sons, Chichester.

3 The record of environmental change in ocean sediments

3.1 Introduction

One of the most significant advances in the elucidation of the natural environmental change of the last 3×10^6 years was the inception of research on ocean-sediment cores in the 1950s (Section 1.2.2). Since then, a wealth of information has been obtained from a large range of cores obtained from such diverse locations as the Antarctic Ocean and the Caribbean. Moreover, in the last 40 years a range of techniques has been devised to extract the physical, chemical and biological secrets concealed within the oceans' sediments.

The major contribution of ocean-sediment studies to environmental-change research lies in their capacity to provide an unbroken record for the last 3×10^6 years. This contrasts with the fragmented nature of the terrestrial record of environmental change. The availability of long sequences of palaeoenvironmental information from various parts of the world is now expediting the establishment of a global framework for environmental change. In particular, the use of ratios of isotopes of oxygen and their dated changes through time have allowed the identification of **chronostratigraphic units**, i.e. firmly dated units, which are referred to as **oxygen isotope stages** or **marine isotope stages**. In order to establish regional and global relationships between disparate deposits, these deposits are increasingly being referred to, and correlated with, marine isotope stages. As well as this framework function, data from ocean-sediment cores have been used to reconstruct sea-surface temperatures, the degree of aridity/humidity on the continents, changes in global ice volume and glacial/interglacial reorganisation of the global carbon cycle.

3.2 Theory and methodology

The oceans currently receive between 6×10^9 and 11×10^9 tonnes of sediment annually. Whilst the volume would have varied in the past, the volume of sediment entering the ocean basins would have been considerable as the continents were eroded as part of the ongoing cycle of erosion, deposition and uplift (Section 1.2.3). The material entering the ocean basins has two components: a sedimentary or mineral matrix, which may be described as **terrigenous** since it derives from the land, and **biogenic material** (i.e. material derived from organisms) comprising the remains of marine organisms and pollen grains derived from continental vegetation communities.

Ocean sediments provide a proxy record of environmental change, since the components of the sedimentary sequences reflect the conditions prevailing within the oceans and on the nearby continents at the time of deposition. Both the terrigenous and biogenic components of ocean sediments have provided, through the application of a range of techniques, a huge volume of information on the natural environmental changes that have occurred over the last 3×10^6 years and even earlier. The following sections describe the various parameters that have been analysed, and examples are given to illustrate the contribution that each has made to the understanding of environmental change.

3.2.1 Oxygen isotope stratigraphy

The material accumulating in the ocean basins is highly varied. In particular, it includes the remains of marine organisms such as foraminifera, diatoms, coccolithophores and radiolaria (Sections 3.2.2, 3.2.3, 3.2.4 and 3.2.5). These organisms are made up of substances that incorporate oxygen. Foraminifera and coccolithophores, for example, consist of calcareous exoskeletons (Sections 3.2.2 and 3.2.5) of calcium carbonate, whilst diatoms and radiolaria (Sections 3.2.3 and 3.2.4) consist of opaline silica. The examination of variations in the ratio of two isotopes of oxygen, ^{16}O and ^{18}O, in ocean sediments, which was first undertaken in the 1940s and 1950s (Section 1.2.2), has made major contributions to palaeoenvironmental studies. First, variations in oxygen isotope ratios have been used to reconstruct changes in ocean temperatures over the last 3×10^6 years. These, in turn, relate to global ice volumes and sea-level changes. Second, there are rhythmical variations in the ratio that are known to occur world-wide, and which are caused by the variations in climate predicted by the Milankovitch hypothesis (Box 2.2). This temporal correspondence has facilitated the development of a stratigraphic framework that is globally applicable and thus provides a means of correlating distant marine cores. The recent identification of oxygen isotope stratigraphy in ice cores and lacustrine sequences is now extending its capacity for correlation between continental, polar and ocean deposits. The establishment of a correspondence between the magnetic stratigraphy of extensive continental loess deposits (e.g. in China) and marine oxygen isotope stages is also contributing to global change frameworks. In Figure 2.4, for example, the continental sequences of the last 900×10^3 years are related to marine oxygen isotope stages.

Box 3.1

The principles of oxygen isotope analysis

1 The carbonate secreted in the calcareous shells of marine organisms such as foraminifera contains two isotopes of oxygen : ^{18}O and ^{16}O. The ratio in which these occur depends on water temperature and the average is approximately 1 ^{18}O:500 ^{16}O. Similarly, ice caps comprise water with a proportion of ^{18}O. Oxygen-isotope stratigraphy is thus characteristic of both fossil marine organisms such as foraminifera and polar ice cores (Chapter 4).

2 The ratios are measured in relation to a standard, and are recorded as the deviation of ^{18}O from this standard. For carbonate shells, the standard is derived from PDB, a Cretaceous belemnite from the Pee Dee Formation in North Carolina, USA. For ice, water and snow, the standard is derived from Standard Mean Ocean Water (SMOW).

3 Mass spectroscopy is used to determine the volumes of each isotope in a given sample. Oxygen isotope ratios are calculated as positive or negative values relative to the standard, using the following equation:

$$\delta^{18}O = 1000 \times \frac{^{18}O/^{16}O \text{ sample} - ^{18}O/^{16}O \text{ standard}}{^{18}O/^{16}O \text{ standard}}$$

4 The deviation from the standard is an indicator of temperature and global ice volume. This is because fractionation occurs between the two isotopes. The ^{16}O is the lighter of the two, so that water containing ^{16}O ($H_2^{16}O$) is preferentially evaporated; this process is temperature-dependent. The more $H_2^{16}O$ that evaporates the more the remaining seawater becomes $H_2^{18}O$-enriched. The $^{16}O/^{18}O$ characteristics of the seawater are reflected in the carbonate of the foraminifera shells. Moreover, winds carry $H_2^{16}O$-enriched water vapour from the oceans to the poles, where it eventually becomes trapped as ice. Consequently, the ice contains an oxygen isotope signal that is the mirror image of that in ocean sediments.

5 During ice ages, $H_2^{18}O$ became enriched in the oceans and hence in the foraminifera while the ice caps became enriched in $H_2^{16}O$. During interglacials, much of the $H_2^{16}O$ stored in the ice was released, diluting the $H_2^{18}O$ in the oceans and causing a shift in the ratio. Variations in the ratio for the last 3×10^6 years are illustrated in Figure 3.2.

The theory underpinning the use of oxygen isotope ratios is complex, as shown in Box 3.1. Although there is some scepticism surrounding how reliably the ratio variations accurately reflect the temperature changes, oxygen isotope analysis provides considerable insight into past environmental change. Figure 3.1 gives an example of oxygen isotope stratigraphy representing the entire Quaternary period and the upper part of the Tertiary period. The substantial oscillations that occur in the ratio reflect the dynamism of global climatic change. In accord with Emiliani's (1955) original suggestion, the oxygen isotope stages are numbered, with odd numbers representing warm stages and even numbers representing cold

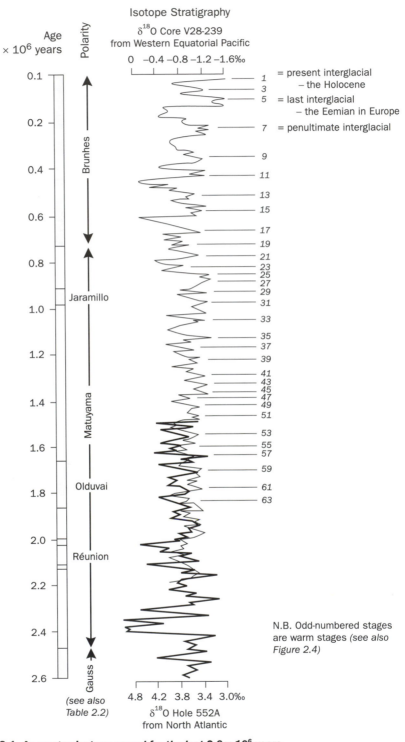

Figure 3.1 *An oxygen isotope record for the last 2.6 × 10⁶ years*

Source: Based on Williams *et al.* (1988) and Shackleton *et al.* (1984).

stages. The rise and fall of the ratio reflects the oscillation of warm and cold stages throughout the period represented by this core from the Pacific Ocean. Clearly, on the time scale represented there is no evidence of climatic 'stability', though there is a high degree of regularity in the oscillations. On the basis of this, and many other ocean-sediment records, there is no reason to suppose that these warm/cold oscillations will not continue into the future.

Other characteristics of environmental change highlighted in Figures 3.1 and 3.2 include the sharp changes that occur at the transitions from cold to warm (glacial to interglacial) stages. For example, between stages 12 and 11 (glacial 5 and interglacial 4 of Figure 2.4), and so on, i.e. stages 10 and 9, 8 and 7, 6 and 5, and 2 and 1 (Figure 2.4) the ratio curve rises rapidly. This reflects a relatively rapid shift from cold to warm conditions; therefore once temperatures begin to increase they do so rapidly. The lesson from this record for future environmental change is that climatic change need not be slow. Indeed, further evidence for the last glacial–interglacial transition (stage 2–1; 14K to 9K years BP), based on fossil coleopteran (beetle) data, indicates that average July temperatures increased by 1°C per decade c. 13K years BP. Thus the term **'termination'**, applied by Broecker and van Donk (1970) to such transitional stages is apt. Several such terminations, as shown in Figure 3.2, have been identified in ocean sediments.

So-called **Heinrich events** have also been recorded in oxygen isotope stratigraphy. Such events, named after their discoverer (Heinrich, 1988), cause the deposition of sediments with abundant **ice-rafted debris** i.e. sedimentary material derived from glacial processes. This can be identified by its sedimentary characteristics (Section 3.2.10) and is considered to represent the increased discharge of icebergs from ice sheets during periods of cooling. Rasmussen et al. (1997) have identified changes in oxygen isotope stratigraphy in cores from the Faeroe–Shetland Channel,

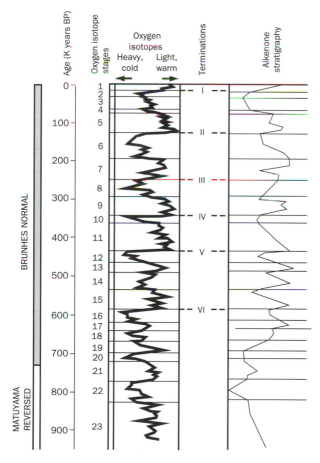

Figure 3.2 *The relationships between oxygen isotope stratigraphy (based on Shackleton and Opdyke, 1973), terminations (Broecker and Van Donk, 1970) and alkenone stratigraphy (Brassell et al., 1986)*

which correlate with Heinrich events and which reflect cooling during the last cold stage. Variations of the oxygen isotope ratios within the major stages illustrated in Figure 3.2 also indicate that they were not internally uniform in relation to temperature (and probably overall climate). For example, the variations in marine oxygen isotope stage 5 (Figure 2.4) represent an interglacial period (stage 5e) and four early stages in the subsequent (last) glacial period (stages 5a, b, c, d).

The value of oxygen isotopes obtained from organisms other than foraminifera for palaeoenvironmental reconstruction has only recently been recognised. Examples will be discussed in the sections below, which consider the role of diatoms, radiolaria, etc. in studies of environmental change.

3.2.2 Foraminifera

Foraminifera are marine protozoa that are unicellular and have a calcareous shell (Figure 3.3), also known as a test. This may comprise a single chamber, or several chambers that are connected via small apertures known as foramina. In many species the chambers are arranged in a spiral, so that the shell is coiled. The largest species may be as much as 100 mm in width and are **benthic**, i.e. bottom dwellers inhabiting the surface or the top few centimetres of the ocean bed. This is in contrast to the **planktonic species**, i.e. free-swimming organisms that live in the water column, which are generally smaller and may be as little as 0.04 mm in width. The calcareous test is produced through the secretion of calcium carbonate. This, in turn, is produced by the foraminifera absorbing the components of the calcium carbonate from the sea-water in which they live. As is discussed above, the isotopic signals

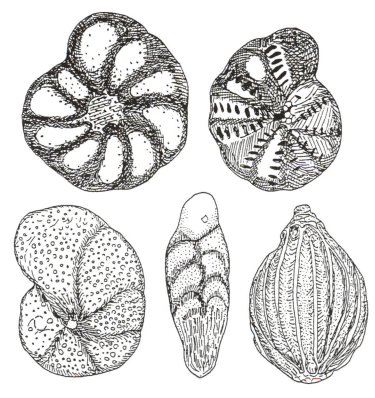

Figure 3.3 *Examples of foraminifera (greatly magnified)*

from the oxygen within the calcium carbonate ($CaCO_3$) of the tests is a particularly significant palaeoenvironmental indicator, because it is related to the prevailing temperatures at the time of formation. At approximately 4 km depth, the calcareous tests dissolve. Consequently, cores for palaeoenvironmental research are usually extracted from regions of the oceans which are less than 4 km deep.

The majority of foraminifera are marine; they occupy a wide range of ecological niches that require a tolerance to salinity. Moreover, their modern-day **biogeography**, i.e. geographical distribution, which reflects a large degree of control by water temperature, is well documented. Consequently, modern analogues provide the means for palaeoenvironmental reconstruction. Indeed, the fact that the geography and ecology of many organisms is well established, is central to palaeoenvironmental reconstructions generally. Most importantly, the temperature regime influences the direction of coiling by foraminiferal tests; many species demonstrate sinistral (to the left) coiling in waters of a certain temperature range, but alter to dextral (to the right) coiling within waters of a significantly lower or higher temperature. This characteristic allows the broad-scale reconstruction of changes in surface-water temperature as the interglacials give way to glacial stages and vice versa. The composition of foraminiferal assemblages has been used to reconstruct the palaeoenvironmental conditions for a wide range of locations. For example, benthic foraminifera from cores extracted from the north-eastern Atlantic Ocean have been used by Thomas *et al.* (1995) to show that during the last glacial maximum *c.* 18K years BP ago, when temperatures were much lower than they are now, the range of species present was relatively low and the accumulation rate of tests within the sediment was also low. This indicates a low productivity rate. As deglaciation occurred *c.* 14K years to 12K years BP, the productivity and species richness increased in response to the increasing temperature.

Foraminiferal assemblages may also be used to estimate sea-level changes. Where such assemblages have been deposited in shallow near-coastal locations, the subsequent sea-level change caused by eustatic and/or isostatic movements (Section 1.2.2) have resulted in such sediments being located above present sea-level. The composition and abundance of the assemblages provides evidence of shallowing and salinity changes. The latter are also indicated where freshwater incursions into the oceans have occurred. For example, Bergsten (1994) has shown that as the last ice age drew to a close, the Skagerrak–Kattegat straits (between Denmark and Sweden/Norway) were dominated by warm, saline water from the Atlantic, notably at 10.2K years BP, whereas earlier the straits received fresh water as the lake occupying the Baltic Basin drained.

3.2.3 Diatoms

Diatoms are unicellular algae, which inhabit all types of aquatic environments including the oceans. They are photosynthetic and form the basis of food chains. The siliceous frustule, or outer shell, has distinctive characteristics, including

shape and pattern, which facilitate identification. Some examples are illustrated in Figure 3.4. Diatoms are particularly common in high-latitude waters, together with radiolaria (Section 3.2.4), whilst organisms with a calcareous skeleton, i.e. coccolithophores (Section 3.2.5) and foraminifera (Section 3.2.2), predominate in low latitudes. Both the diatoms themselves, (in terms of their assemblage characteristics and accumulation rates), and the oxygen isotope signatures that they contain, provide valuable palaeoenvironmental insights.

Figure 3.4 *Examples of diatoms (greatly magnified)*

Williams *et al.* (1995) have used a variety of palaeoenvironmental indicators, including diatoms, from a core from the south-east Greenland Shelf to show that cool/warm pulses occurred on the outer shelf between 13.5K years BP, and 9K years BP and between 9K years BP and 6K years BP on the inner shelf. These water-temperature oscillations were caused by pulses of meltwater from the retreating ice. Moreover, Shemesh *et al.* (1995) have shown that oxygen isotope signatures obtained from diatoms in eight cores from the Atlantic sector of the Southern Ocean consistently reflect similar patterns of sea-surface temperature change *c.* 20K years BP, and 8K years BP, i.e. during the last glacial maximum and the early Holocene. Fossil marine diatom assemblages have also been used to indicate rates of primary productivity during the last glacial/interglacial cycle. Indeed, Pollock (1997) has suggested that increased diatom productivity, stimulated by interglacial warming that causes ice-sheet melting, may ultimately result in global cooling. This may occur because melting ice sheets release silicate-rich water, which is essential for diatom growth. As growth continues atmospheric carbon dioxide is depleted, a process which is accelerated by sea-level fall, causing nutrient release through the erosion of exposed sediments. Whilst this is an interesting possibility, and one which would explain the reduction of atmospheric carbon dioxide during cold stages, there is little evidence available so far to support it. As more diatom data from ocean cores become available, this hypothesis can be tested.

3.2.4 Radiolaria

Like diatoms, radiolaria have skeletons composed of opaline silica (Figure 3.5).

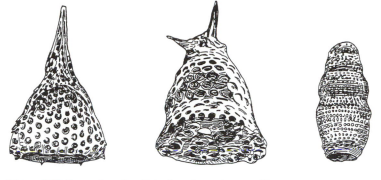

They are single-celled amoebic protozoa that are **heterotrophic organisms**, i.e. they do not photosynthesise. Radiolarian assemblages reflect changes in salinity, temperature and water depth. The few studies so far undertaken on radiolaria in a palaeo-oceanic context indicate

Figure 3.5 *Examples of radiolarians (greatly magnified)*

that they have considerable potential as palaeoenvironmental indicators.

The role of radiolaria as indicators of salinity change has been illustrated by the work of Gupta and Fernandes (1997) on a core from the tropical Indian Ocean. The reconstructed changes in salinity for the period 1400K years BP to 200K years BP are considered to reflect changes in the palaeomonsoon. The monsoon brings an influx of fresh water via direct precipitation and increased river inflow, which results in reduced salinity. Moreover, Gupta and Fernandes have identified cyclical variations in the palaeo-monsoon, which they believe corresponds to Milankovitch's 100K year variation in orbital eccentricity (Box 2.2). Radiolarian assemblages have also been used to reconstruct fluctuations in the monsoon of the north-west Pacific for the last 350K years (Heusser and Morley, 1997). The stages relating to the interglacials, for example, exhibit periods of between 10K years and 15K years duration which are characterised by enhanced precipitation due to the intensification of the summer monsoon.

The sensitivity of radiolaria to temperature change has been exploited to reconstruct sea-surface temperatures in the north-eastern Pacific; these are considered to be controlled by the position of the North Pacific high-pressure cell and its influence on ocean circulation (Sabin and Pisias, 1996). Radiolarian assemblages from eight cores extracted from the eastern equatorial Pacific have also been used to show that sea-surface temperatures were 3 to 5°C lower during the last glacial maximum than they are now (Pisias and Mix, 1997).

As in the case of diatoms, research into the palaeoenvironmental value of radiolarians is only just beginning, and their full potential has yet to be realised.

3.2.5 Coccolithophores

Coccolithophores are marine calcareous **nanoplankton**. This means that they are very small organisms of less than 100 µm in diameter. Like the diatoms, coccolithophores are **primary producers**, i.e. autotrophic organisms that photosynthesise and thus form the basis of food chains and webs. Unlike the diatoms, they possess a skeleton of calcium carbonate, which is precipitated as calcite. The calcite is deposited in species-specific patterns, as illustrated in Figure 3.6; this facilitates identification. Coccolithophores are abundant and are a

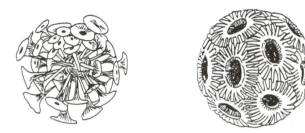

major constituent of the calcareous oozes on the ocean bed. They are sensitive to salinity and are most abundant at salinities of between 38‰ and 25‰. The intensity of light is also a limiting factor, but the most significant control on their distribution and abundance is temperature. It is this characteristic that has proved to be particularly

Figure 3.6 *Examples of coccolithophores (greatly magnified)*

valuable in palaeo-oceanic research. Interest has focused on assemblage composition to determine change in surface-water temperatures and, more recently, attention has turned to examining variations in the composition of complex organic chemical substances known as alkenones.

In relation to species distribution, Weaver and Pujol (1988) have examined the change in species composition that occurred in cores from the Alboran Sea (western Mediterranean) during the period 15K years BP to 10K years BP. This is an especially complex period in terms of environmental change (Chapter 5), when the last ice age drew to a close in two stages. It involved an initial termination, (Section 3.2.1) separated from a second termination by a warm period (equivalent to the Bølling/Allerød stage discussed in Chapter 5). The Alboran coccolith assemblages reflect these changes, with *Geophyrocapsa muellerae*, a relatively cold-tolerant species, dominating the sediments prior to each termination, whilst a warmth-demanding species, *G. oceania*, was most abundant in the sediments deposited during the intervening warm stage. In another study from the western Mediterranean, covering oxygen isotope stages 5 to 1, it was found that coccolithophore concentrations increased markedly during warm periods, especially in interglacials, when compared with the glacial stages, which were also characterised by an abundance of reworked specimens (Flores *et al.*, 1997).

However, perhaps the most exciting aspect of research on coccolithophores is the prospect of precise reconstruction of sea-surface temperatures afforded by their **alkenone** (lipid components of cell membranes) content. This possibility was first highlighted by Brassell *et al.* (1986), who demonstrated that alkenones of the coccolithophore *Emiliania huxleyi* varied in terms of the degree of saturation

(Box 3.2), depending on their geographical distribution. They then showed that the alkenone characteristics varied within the sediments of a core from the Kane Gap in the eastern equatorial Atlantic, and that these variations paralleled changes in oxygen isotope signatures, which are also related to temperature (Section 3.2.1). On the basis of these observations, Brassell *et al.* suggested the formulation of the U_{37}^{k} index i.e. the **alkenone unsaturation index**. In essence, cold stages are characterised by a low U_{37}^{k} index, whilst warm stages are characterised by a high U_{37}^{k} index (Box 3.2; Figure 3.2). This change in chemical structure reflects the capacity of the organism to adapt to temperature change, and thus may provide a more reliable indicator of past sea-surface temperatures than oxygen isotope signatures, since these are also influenced by other variables.

Examples of the application of alkenone stratigraphy, i.e. variations in the U_{37}^{k} index, include the work of Emeis *et al.* (1995), and Schneider *et al.* (1995), on cores from the Arabian Sea and the equatorial South Atlantic respectively. The former core represents the past 500K years, and temperature reconstructions show that interglacial stage sea-surface temperatures were as high as 27°C, but declined by between 2°C and 5°C during glacial stages. However, recent research has shown that there is a substantial discrepancy between sea-surface temperature reconstructions based on alkenones and those based on planktonic foraminiferal assemblages. For example, Holocene temperature reconstructions for an eastern

---

Box 3.2

The principles of alkenone stratigraphy

1 Alkenones are components of lipids (fats) in cells of marine organisms, notably coccolithophores which are calcareous plankton.

2 Alkenones have a large number of carbon atoms, usually C_{37} to C_{39}, which vary in their degree of saturation. This refers to the nature of the chemical bonds between the carbon atoms. Unsaturated molecules are characterised by single rather than multiple bonds between the carbon atoms. Most importantly, the degree of unsaturation, at least in the coccolithophore *Emiliani huxleyi,* is considered to be temperature dependent (Brassell *et al.*, 1986).

3 Unsaturation prevails in organisms living in relatively warm conditions while saturation prevails in organisms living in relatively cold conditions. This relationship has been observed in the laboratory and in terms of the modern biogeography of *Emiliani huxleyi* i.e. in northern ocean waters, modern specimens exhibited a lower degree of unsaturation than specimens in tropical waters.

4 The use of an unsaturated alkenone index, U_{37}^{k}, may provide a means of reconstructing sea-surface temperatures and one which is independent of oxygen isotope signatures. An example is given in Figure 3.2.

subtropical Atlantic core vary by 3°C for 8K years BP (Chapman *et al.*, 1996). Research on a north-east Atlantic core revealed even larger discrepancies between the results of the two methods of temperature reconstruction. Foraminiferal data reflect a sea-surface temperature of 9°C for the early stages of deglaciation and 13°C for the Holocene; this is in contrast to 13°C and 18°C derived from alkenone data. Conversely, however, Pisias *et al.* (1997) have found good agreement between sea-surface temperature reconstructions based on radiolaria, foraminifera and U_{37}^k indices for the last 20K years from a north-east Pacific core. Clearly, further research is required to reconcile such variations.

3.2.6 Marine molluscs

There are many different types of marine molluscs, which occupy a wide range of habitats, ranging from coastal pools to the sediments of the continental shelf. Analyses of their remains, which are usually identified by the shape and pattern of their calcareous shells, have been undertaken from sediments derived from all these environments, including sediments originally deposited in coastal situations but now firmly on land. This reflects sea-level change, and the extent and nature of such sediments and their fossils provide evidence for the magnitude of this type of change.

As in the case of the other organisms referred to in this chapter, palaeo-environmental reconstructions are based on the present-day distributions and ecological preferences of individual species. In addition, the chemical components of molluscan shells are beginning to be exploited as palaeoenvironmental indicators.

One example of the value of marine mollusc remains as indicators of sea-level change is given by the work of Clapperton *et al.* (1995) on the central Magellan Strait in southern Chile. Here, during oxygen isotope stages 5b to 1 representing the last glacial cycle (Figure 3.2) the presence of marine mollusc shells in basal tills indicates the occurrence of several marine incursions between periods of ice regression. Not only does this study facilitate the reconstruction of local events, but it also illustrates the fact that considerable environmental change occurs during ice ages. The modern biogeography of marine molluscs has been used in a study to reconstruct environmental change during the last 18K years in the North Atlantic (Dyke *et al.*, 1996). The study shows that, as deglaciation occurred, the extent of the arctic zone declined. At the last glacial maximum, *c.* 18K years BP, it extended from the Arctic Ocean to the Grand Banks, but as the climate ameliorated the biogeographical zones moved north.

Oxygen isotope signatures from marine mollusc remains in an archaeological cave deposit on the coast of Agulhas Bank, southern Africa, have also been used to reconstruct sea-surface temperatures (Cohen and Tyson, 1995). In combination with radiocarbon dating directly on the shells, it was determined that, during the early Holocene, sea-surface temperatures adjacent to the coast were lower than they are today, but that, by *c.* 5.8K years BP, summer and winter sea-surface temperatures were higher by *c.* 2°C. The significance of molluscs as indicators of palaeosalinities is illustrated by the work of Holmden *et al.* (1997), who have employed the ratios of strontium-87 to strontium-86. Based on characteristic strontium ratios in modern shells from marine and fluvial (i.e. freshwater) environments, the reconstruction of past salinities is feasible. The examples given by Holmden *et al.* involve the reconstruction of early Cretaceous salinities, but the underlying premises could be applied to deposits of the last 3×10^6 years. Other studies have involved the use of individual indicator species, as discussed in Lowe and Walker (1997).

3.2.7 Ostracods

The Ostracoda are a class of microscopic crustacea with bivalve shells. In common with the molluscs (Section 3.2.6), many ostracods have calcitic shells, whilst others have **chitinous** shells (chitin is a complex organic substance that forms the exoskeletons of insects). They are multicellular organisms, mostly in the size range 0.6 to 2 mm. Modern-day ostracods occupy aquatic ecological niches that extend across a wide range of salinities. In addition, their productivity is temperature dependent.

The latter characteristic has been used to determine changes in productivity in the Arctic Ocean during the last 300K years (Cronin *et al.*, 1994). In particular, productivity increased substantially during oxygen isotope stages 1, 5 and 7, i.e. the present and last two interglacial stages. The value of ostracod assemblages for reconstructing salinity changes has been demonstrated by the work of Andren and Sohlenius (1995) on the north-west Baltic Sea area. Based on the analyses of seven cores, they were able to identify a marine phase that occurred at the end of the last ice age when the Baltic Sea was a freshwater lake. Similarly, Aguirre and Whatley (1995) have shown that the transition period between the last glacial stage and the Holocene in the south-western Atlantic, off the coast of Buenos Aires, was characterised by cool freshwater conditions rather than marine conditions. This indicates that sea-level was lower at this time, *c.* 11K years BP, than during the later Holocene.

3.2.8 Pollen

Along the continental margins the accumulating marine sediments also receive pollen from the vegetation communities of adjacent continental landmasses.

Consequently, these sediments contain a record of continental vegetation change, which will reflect climatic change, as well as a record of marine changes. Pollen grains can be identified to genus level, and pollen assemblages provide valuable insights into environmental change, as discussed in Chapter 5 (Box 5.1).

The value of pollen assemblages from marine sediments is illustrated by the work of Heusser and Morley (1997), which is referred to in Section 3.2.4 in relation to radiolaria. Their reconstruction of climatic change for the last three glacial–interglacial cycles in the north-west Pacific, off the coast of eastern Japan, indicates that each interglacial was quite distinct. This reflects variability in terms of the sequence, timing and intensity of climatic change. Moreover, the occurrence of high frequencies of the pollen of *Cryptomeria*, the Japanese cedar, in all the interglacial periods, is considered to reflect increased precipitation brought by an intensifying monsoon. Pollen assemblages from a core from the Angola Basin have been used by Ning and Dupont (1997) to reconstruct shifts in vegetation belts which have occurred during the last 300K years on the adjacent continent. For example, glacial stages were characterised by high percentages of desert and semi-desert plants, reflecting the dominance of arid conditions on the continent.

3.2.9 Indices of productivity

There are several indices of productivity other than the accumulation rates of marine organisms. Marine microfossils contain trace elements, the quantity of which reflect abundance in the ocean water at the time of uptake. Some trace elements, e.g. cadmium and barium, are considered to reflect nutrient concentrations in ocean waters; these in turn influence productivity. Cadmium is similar to phosphorus in relation to its circulation in ocean waters and uptake by organisms. Consequently, cadmium/calcium ratios (Cd/Ca) in marine organisms such as the calcareous foraminifera are considered to reflect past concentrations of phosphorus, an important and vital nutrient (Boyle, 1990). One example of the application of Cd/Ca ratios is that of Keigwin *et al.* (1991). Their analysis of Cd/Ca ratios in foraminifera from a North Atlantic core showed that, as the last ice age ended, at *c.* 14.5K years to 10.5K years BP, the ratio was low during four periods; these are thought to reflect inputs of meltwater, which diluted the amount of cadmium available. Barium concentrations in marine fossils are also considered to reflect productivity, and the germanium concentrations in siliceous fossils such as diatoms and radiolaria are considered to reflect the amount of silica in ocean waters.

3.2.10 Sedimentary characteristics

There are many sedimentary characteristics that provide insights into environmental change. Composition, grain size and roughness/smoothness, for example, provide information on sediment source, mode of transport and wind direction. The latter is often climate dependent, so palaeoclimatological inferences can be drawn. For example, the varied nature of the ice-rafted debris (Section 3.2.1) in sediment cores from the Fram Strait, which connects the Arctic Ocean to the Atlantic, reflects its varied origin (Hebbeln and Wefer, 1997). In oxygen isotope stage 6, ice-rafted material came from Siberia; during the major glacial advances of stages 2 to 5 the debris came from the Svalbard/Barents Sea region; and during short-lived open-water periods within the glacial stages, material from Fennoscandinavia was deposited. Moreover, the presence of several ice-rafted debris layers in sediment cores from the Norwegian Sea has been used as a proxy record of ice advances in the shelf area over the last 150K years (Baumann et al., 1995). Similarly, the presence of Heinrich layers (Section 3.2.1), characterised by a high terrigenous content, has been used to indicate the presence of icebergs during the glacial stages of the last 225K years off the coast of Portugal (Lebreiro et al., 1996). This study suggests that icebergs extended for a considerable distance to the south of the parent ice cap in the North Atlantic.

The composition of clays deposited off the coast of French Guiana has been used to identify source areas, and is related to climatic changes during the last 3000 years. According to Pujos et al. (1996), muds with a high illite and chlorite content derive from the Andes, whilst those with a high kaolinite and smectite content derive from the Amazonian lowlands and the Guiana Shield. Deposition between 3K years BP and 1.7K years BP, and between 1.7K years BP and 1K years BP, derives from the Amazon Lowlands and Guiana. Between 2.2K years BP and 1.2K years BP there was a significant reduction in material from the Andes, which reflects a climatic shift to dryness resulting in a decrease of water erosion.

3.3 Palaeo-oceanic reconstructions

One of the most significant collaborative research projects on palaeoenvironments was the CLIMAP project (Climate Long-ranged Investigation Mapping and Predictions), which was established in the late 1970s (CLIMAP, 1976). It involved the collection and analysis of cores from diverse parts of the world's oceans, and its data and results have made a major contribution to current understanding of past climates. One output from the project is illustrated in Figure 3.7. This comprises two maps depicting reconstructed sea-surface temperatures during the last glacial maximum, c. 18K years BP, and modern sea-surface temperatures. Of particular note is the considerable difference in temperature which occurred in the

A. 18K years BP (last glacial maximum)

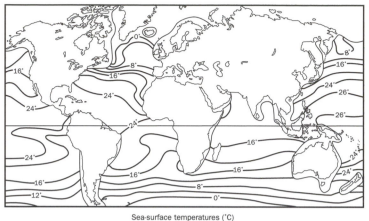

Sea-surface temperatures (˚C)

B. Present day

Figure 3.7 *The reconstruction of sea-surface temperatures in August for the last glacial maximum c. 18K years BP (A) in relation to those of the present day (B)*

Source: Based on CLIMAP (1981).

North Atlantic. The CLIMAP project data are widely used in climate modelling experiments. For example, the predictions of general circulation models can be tested against the CLIMAP results.

3.4 Conclusions

The sediments of the world's oceans contain a wealth of information on the environment of the last 3×10^6 years of the Earth's history. Both terrigenous and biogenic components can be exploited. In particular, the oxygen isotope ratios derived from marine fossils, notably from foraminifera, have facilitated the establishment of a stratigraphic framework, which now has global applications.

Both calcareous fossils, i.e. foraminifera, molluscs, ostracods and coccolithophores and siliceous fossils, i.e. radiolaria and diatoms, provide information on palaeo-productivity, past oceanic biogeography and ocean water characteristics. The characteristics of sediments in marine cores provide data on ice advance, the direction of ice movement and continental environmental change.

Most studies of marine sediments are multidisciplinary and involve a range of techniques along with age determination. Such studies have made a major contribution to the understanding of past environmental change and will continue to do so in the future as new techniques of investigation are developed.

Summary Points

- Selected ocean sediments contain an unbroken record of environmental change over the last 3×10^6 years.

- Oxygen isotope signatures, derived mainly from foraminifera, constitute a valuable stratigraphic tool.

- The **biogenic** and terrestrial components of ocean sediments provide palaeoenvironmental information.

- Calcareous fossils include foraminifera, molluscs, ostracods and coccolithophores.

- Siliceous fossils include radiolaria and diatoms.

- These fossils and their chemical constitution provide information on oceanic circulation, palaeo-productivity, temperature change and salinity.

- The terrigenous component provides information on ice advance, sediment source and meltwater discharge.

- Marine-sediment stratigraphy provides a means of correlation with continental sequences such as lake sediments and loess.

General further reading

Environmental Change. A. Goudie. 1992. Clarendon Press, Oxford, 3rd edn.

Global Environmental Change. A Natural and Cultural Environmental History. A.M. Mannion. 1997. Longman, Harlow, Essex, 2nd edn.

Global Environmental Change. P.D. Moore, B. Chaloner and P. Stott. 1996. Blackwell, Oxford.

Ice Age Earth. A.G. Dawson. 1992. Routledge, London.

Late Quaternary Palaeoceanography of the North Atlantic Margins. J.T. Andrews, W.E.N. Austin, H. Bergsten and A.E. Jennings (eds). 1996. Geological Society Special Publication, London.

Reconstructing Quaternary Environments. J.J. Lowe and M.J.C. Walker. 1997. Longman, Harlow, Essex, 2nd edn.

4 The record of environmental change in ice cores

4.1 Introduction

The first cores from the world's major ice sheets were extracted in the 1960s. Since then, ice cores from a variety of locations have made a major contribution to palaeoenvironmental studies. Of particular note was the raising of a core from Vostok, Antarctica, in the early 1980s, which covered the last glacial/interglacial cycle (last *c.* 160K years). Recently, two projects have led to the extraction of two long cores from the Greenland ice sheet: the Greenland Ice Core Project (GRIP) was organised under the auspices of the European Science Foundation and the Greenland Ice Sheet Project (GISP). The latter is a North American project.

Whilst these three ice cores have altered ideas on environmental change, and generated controversy through apparent lack of correspondence, there are a number of cores from other regions. For example, cores from Peru, Mongolia and Tibet are now available, and analysis of their components is generating valuable data on environmental change. The availability of palaeoenvironmental data from high latitudes and altitudes provides another significant piece in the global jigsaw of environmental change. Historically (Section 1.2), the record of environmental change in terrestrial and lacustrine sequences (Chapter 5) was the first to receive attention. Later, interest in the oceans and their sediments developed (Chapter 3); this, mainly through the establishment of a widely applicable oxygen isotope stratigraphy (Figure 3.1), provided a framework for global environmental change. Ice-core analysis provides yet another facet of environmental change studies, which introduces information from a different set of archives: glaciers and ice sheets.

Numerous indices of environmental change have been developed in order to exploit these frozen archives. Of these innovations, the establishment of an oxygen isotope stratigraphy is particularly important. This is because it provides both an index of temperature change and a means of correlation between ocean sediments, continental sequences and the ice cores. In addition, the ice cores, through the presence of bubbles, offer a unique opportunity to measure directly the composition of the atmosphere and its temporal change.

4.2 Ice-core locations

Although most attention has focused on the long ice cores obtained in recent years from the Arctic and Antarctic, i.e. GRIP, GISP and Vostok, many other cores from these inhospitable environments have been highly significant in terms of understanding the dynamics of environmental change. The locations of these cores are given in Figure 4.1. It should also be noted that these cores have provided valuable information on both naturally driven and culturally driven environmental change. The latter includes a record of various types of pollution.

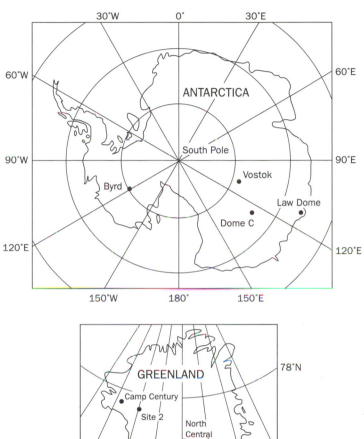

Figure 4.1 *The location of major ice-core sites*

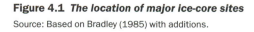

Source: Based on Bradley (1985) with additions.

4.3 Chemical characteristics

Ice cores contain much more than water, though this in itself houses valuable palaeoenvironmental information through the oxygen isotope signatures it contains (Section 3.2.1; Box 3.1). Similarly, the presence of heavy water (deuterium oxide) is an indicator of past temperatures. Other chemical characteristics reflect past volcanic activity and dust input. Of major significance is the presence of bubbles in the ice; these have captured 'mini' atmospher   f the past, making it possible to establish the magnitude and timing of changes   .mospheric composition.

[handwritten: glacial less O¹⁸ - O¹⁶ doesn't make to higher latitudes b/c condense out quicker]
[handwritten: interglc more O¹⁸ — b/c doesn't condus]

4.3.1 Oxygen isotope stratigraphy

The interpretation of oxygen isotope signatures in marine cores was discussed in Section 3.2.1 and Box 3.1. Essentially, water evaporated from the oceans becomes enriched in the lighter isotope ^{16}O and depleted in the heavier isotope ^{18}O. During ice ages/cold stages, the ice sheets thus became enriched in $H_2{}^{16}O$, while the oceans became enriched in $H_2{}^{18}O$. Figure 4.2 gives oxygen isotope data from the Vostok core, which extends back to c. 160K years BP. Clearly, there are major variations, which have been used by Lorius et al. (1985) to subdivide the record into eight stages. Stage G represents the last **interglacial**, while A represents the Holocene, i.e. the present interglacial. Stage H is the terminal part of the penultimate glacial period, while stages B to F represent the last **glacial** period. The non-uniformity of this glacial period is apparent and it is considered that stages C and E reflect **interstadials**, while stages B, D and F reflect **stadials** (see Section 2.3 and Glossary for definitions). The magnitude of change between stages H and G, for example, reflects a sharp and relatively rapid shift from a glacial to an interglacial state.

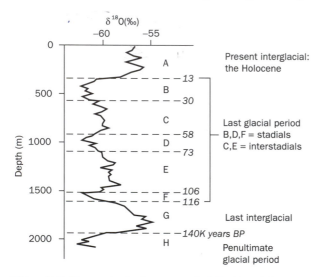

Figure 4.2 **The oxygen isotope record of the Vostok ice core, Antarctica**

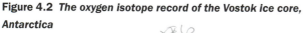

Source: Based on Lorius et al. (1985). *[handwritten: p86]*

This is repeated at the transition between stage A and B. Such changes involved a change in temperature of between 8 and 10°C, whilst the change between stadials and interstadials was probably between 2 and 4°C (Lorius et al., 1985).

More recently, the oxygen isotope signatures from the GRIP and GISP2 cores have been published. These have created much controversy because, as shown in Figure 4.3, there is considerable variation between the two in the section close to the base of the cores, which represents the last interglacial. The controversy centres on the fact that in the GRIP core the oxygen isotope concentrations are erratic, whilst those in the GISP2 core are not. Thus the GRIP data may reflect an unstable climate, whilst the GISP2 data indicate a stable climate. The close proximity of the locations of these cores makes this discrepancy particularly

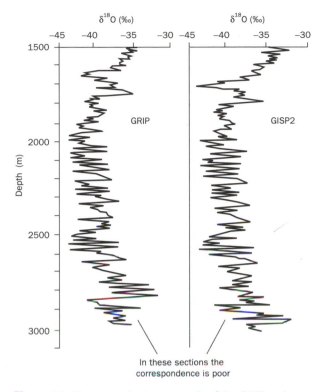

Figure 4.3 *The oxygen isotope records of the GRIP and GISP2 (Greenland) ice cores*

Source: Based on Grootes *et al.* (1993).

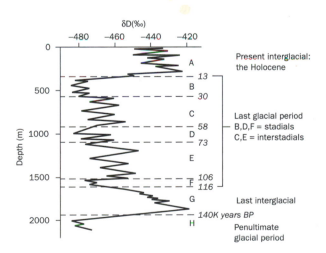

Figure 4.4 *The deuterium record from the Vostok ice core, Antarctica*

Source: Based on Jouzel *et al.* (1987).

remarkable. However, evidence from elsewhere generally implies stability, e.g. from marine sediments (McManus *et al.*, 1994), and from marine sediments compared with pollen assemblages (Kukla *et al.*, 1997). The issue is not yet fully resolved.

4.3.2 Heavy water (deuterium oxide)

A proportion of the water deposited in ice caps and glaciers contains a heavy isotope of hydrogen, i.e. ^{2}H, while most of the water deposited comprises ^{1}H. The fact that ^{2}H is heavier than ^{1}H means that it behaves like the heavy isotope of oxygen (^{18}O). Consequently, the deuterium profile in an ice core is an additional proxy record of temperature change. The deuterium record for the Vostok core is given in Figure 4.4, and this record has been subdivided into eight stages on the basis of the oxygen isotope stratigraphy (Section 4.3.1). The deuterium curve mirrors that of the oxygen isotope curve of Figure 4.2. Moreover, both the deuterium and oxygen isotope records indicate that the last interglacial was *c.* 2°C warmer than the present interglacial (Jouzel *et al.*, 1987).

Recently, another ice core has been obtained from Vostok. This has extended the temporal record beyond the original 160K year BP to *c.* 260K years BP. The deuterium profile from this new core has led to the suggestion that during the marine oxygen isotope stage 7 (a warm period; see Figure 2.4) temperatures in Antarctica were as warm as those of today, and that the temperature regimes of the last two climatic cycles (stages 6 and 7, stages 1–5) were very similar (Jouzel *et al.*, 1996).

4.3.3 Carbon dioxide and methane

The extraction of the Vostok core was not only a technological feat, but also

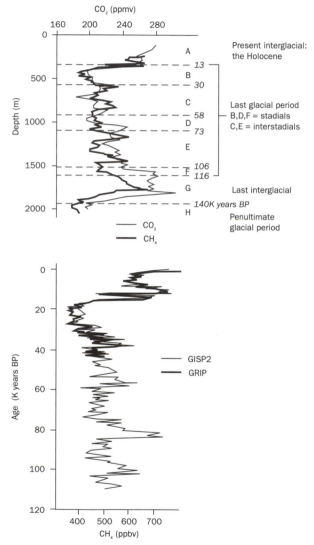

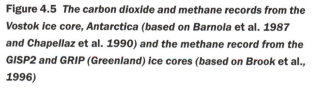

provided a means of directly recording the changing atmospheric composition of the last climatic cycle. This is possible because the bubbles in the ice provide a temporal record of atmospheric composition; as snow layers accumulate and are compressed into ice, bubbles form, and these encapsulate a mini atmosphere representative of the atmosphere at that time. Figure 4.5 illustrates how the concentrations of carbon dioxide (Barnola *et al.*, 1987) and methane (Raynaud *et al.*, 1988) have changed during the last 160K years. These two gases are particularly significant because they are heat-trapping gases (this is why so much concern is currently being expressed about their increasing release into the atmosphere and global warming), i.e. they are greenhouse gases. Moreover, they are important components of the global carbon cycle, which has altered in tandem with the accumulation and melting of the ice sheets. Changes in the concentrations of these gases thus provide information on the intensity of the greenhouse effect and the significance of the atmospheric pool of carbon.

Figure 4.5 *The carbon dioxide and methane records from the Vostok ice core, Antarctica (based on Barnola et al. 1987 and Chapellaz et al. 1990) and the methane record from the GISP2 and GRIP (Greenland) ice cores (based on Brook et al., 1996)*

One of the most marked features of Figure 4.5 is the parallelism between the trends for carbon dioxide and methane. The data also show distinct differences between the glacial stages

(H and B, D and F) and the interglacial (A, G) and interstadial (C, E) stages. The onset of an interglacial, for example between stages B and A or between stages H and G, is characterised by rapid and substantial increases in both of these gases. The concentration of carbon dioxide increases by *c.* 25 per cent, whilst the concentration of methane doubles. Similar trends for methane occur in the Greenland cores (Brook *et al.*, 1996), as illustrated in Figure 4.5.

These data confirm that there is an important relationship between climatic change and the global carbon cycle. Moreover, the significance of the Milankovitch cycles in climatic change (Section 2.4, Box 2.2) implies that there is a link between these and changes in the pools and fluxes of the carbon cycle. This is a controversial issue insofar as it is not clear whether changes in the carbon cycle were **forcing** or **reinforcing factors** in climatic change, i.e. such changes may have initiated positive **feedback** or they may have resulted from the stimulus of an external factor such as astronomical forcing. Part of this problem of resolution is caused by difficulties in estimating age, notably the margins of error associated with techniques such as radiocarbon dating. Nevertheless, the close involvement of the carbon cycle in glacial/interglacial climatic cycles also implies that organisms are involved; this is the essence of the **Gaia** hypothesis, which espouses the view that life influences and is influenced by atmospheric composition (Section 1.2.4). Although there are numerous suggestions as to how this might come about, there is no general agreement (Section 4.3.7; Mannion, 1997b).

4.3.4 Sodium, aluminium and other heavy metals

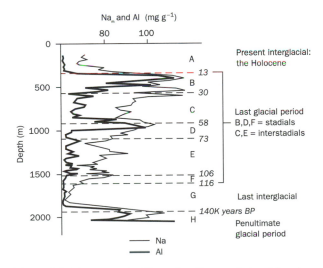

Figure 4.6 *The concentrations of sodium and aluminium in the Vostok ice core, Antarctica*

Source: Based on De Angelis *et al.*, (1987).

Figure 4.6 illustrates the variations in sodium and aluminium that characterise the Vostok core. Sodium is considered to reflect the volume of inputs to Antarctic ice from marine sources, e.g. the salts sodium chloride and sodium sulphate. Variations in the concentrations of aluminium reflect the magnitude of dust input derived from the nearby continents, and are considered to reflect the degree of continental aridity. For example, Figure 4.6 shows that during the last and penultimate glacial stages, the concentrations of both sodium and aluminium were high, if erratic, when compared with interglacial stages. In terms of palaeoenvironmental reconstruction, these variations are thought to reflect the persistence of arid continental conditions during the glacial stages, and probably the exposure through lowered sea-levels, of continental shelf areas from which dust was blown on to Antarctica. Analyses of heavy metals in the GRIP core also show a consistent variation between glacial and interglacial periods. According to Hong et al. (1996), the high concentrations during glacial times were due to intense erosion in Eurasia and North America. High sodium concentrations reflect increased inputs from the ocean to the ice sheet, which may reflect the pattern of wind circulation. Analysis of the dust itself in ice cores can also help isolate the source area of wind-blown material, as discussed in Section 4.4.2.

4.3.5 Beryllium (^{10}Be)

Variations in beryllium concentrations in ice cores, notably Vostok, have also received attention. For example, there are parallels between variations in ^{10}Be and ^{14}C (radiocarbon). Thus the Vostok ^{10}Be record, illustrated in Figure 4.7, may be a proxy record for ^{14}C. This has implications for the use of ^{14}C as a means of age determination for late Quaternary materials, as errors are generated by assuming a constant rate of ^{14}C occurrence. The ^{10}Be record may also be related to solar and geomagnetic changes, as discussed by Stuiver and Brazunias (1993). In addition, the ^{10}Be record has been considered as a proxy for precipitation. Although this relationship is tenuous, Yiou et al. (1985) have suggested that the record illustrated in Figure 4.7 indicates precipitation concentrations of c. 50 per cent less during glacial stages as compared with the current or last interglacial stages. This does correspond with aluminium records (Section 4.3.4; Figure 4.6) and dust records (Section 4.4.2) for enhanced continental aridity during glacial stages.

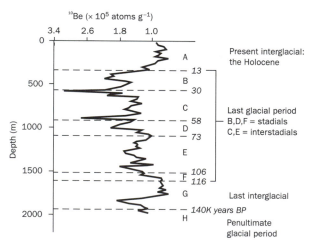

Figure 4.7 *The record of beryllium (^{10}Be) from the Vostok ice core, Antarctica*

Source: Based on Raisbeck et al., (1987).

4.3.6 Acidity (anions)

Acidity variations in ice cores, as reflected in the concentrations of anions, including nitrate and sulphate, are considered to reflect periods of volcanic activity when acidic aerosols are produced in abundance. These, in turn, are often invoked as agents of climatic change. Figure 4.8, the acidity record of the Vostok core, shows there is no relationship with any of the variables discussed above. Although the peaks can often be correlated with historically documented or sedimentologically recognised and dated volcanic eruptions of some magnitude, the lack of correlation between acidity and most other indices of major climatic change, i.e. a glacial/interglacial cycle, implies that volcanic eruptions have no long-term effect on global climates. A further implication is that volcanic activity is not a direct cause of glacial/interglacial cycles.

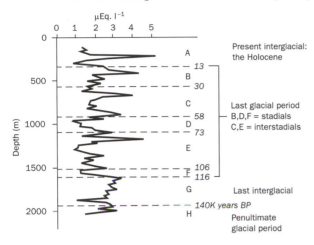

Figure 4.8 *The acidity profile of the Vostok ice core, Antarctica*

Source: Based on Legrand *et al.* (1988).

However, the record of acidity in ice cores is valuable, particularly because it provides a means of correlation with ice cores from elsewhere, and also because it allows the establishment of a chronology of significant volcanic activity which predates written records. This is exemplified by the work of Zielinski *et al.* (1996) who have produced a record of volcanism for the last 100K years from sulphate concentrations in the GISP2 core. This record shows that the eruptions were concentrated during two specific time periods: 17K to 6K years BP and 35K to 22K years BP. Overall, Zielinski *et al.* have identified 850 volcanic signals, though there is no unequivocal evidence that such eruptions have caused climatic change. In a recent review of sulphate records in ice cores, Legrand (1997) has noted that in the Antarctic the sulphate is mostly of biogenic origin (Section 4.3.7), while in Greenland, sulphate concentrations have increased by a factor of four since *c.* 1900 as a result of fossil-fuel combustion. Such records, discussed again in Section 4.4.3 in relation to volcanic ash, also provide markers that facilitate correlation between disparate sequences. Electrical conductivity measurements (ECM) and dielectric properties (DEP) can also be determined for ice cores. Although these properties are essentially physical rather than chemical characteristics they reflect the concentrations of acids and hence provide a proxy record of volcanic activity.

4.3.7 Methanesulphonate (MSA)

The sulphate referred to above derives from volcanic aerosols and is an important component of the global sulphur cycle. As Lovelock (1997) has discussed, biological (**biogenic**) emissions of sulphur also constitute important fluxes between the biospheric and atmospheric pools of sulphur. Such emissions include dimethyl sulphide (DMS – $(CH_3)_2S$), which is produced by algae; the oxidation product of DMS, i.e. MSA, is recorded in polar ice cores. The Vostok record of MSA is given in Figure 4.9, which shows that MSA concentrations are highest during ice ages and lowest during interglacials. The implication of this trend is that DMS production, and hence algal activity, was higher during ice ages. This may relate to increased marine primary productivity, which is postulated to have contributed to a diminution of the greenhouse effect during ice ages, i.e. increased marine biotic activity results in a drawdown of carbon dioxide and so contributes to global cooling.

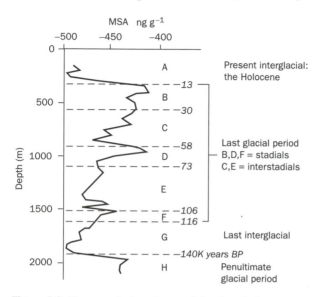

Figure 4.9 *The record of methanesulphonic acid from the Vostok ice core, Antarctica*

Source: Based on Legrand (1997).

Indeed it has been argued (e.g. Pollock, 1997; Section 3.2.3) that such productivity may not just contribute to global cooling, but may also constitute a prime cause, with stimulation to algal growth coming from increased nutrient inputs due to the presence of wind-blown material. In particular, attention has focused on the role of iron in marine productivity and its possible increase during ice ages (Martin, 1990; Martin *et al.*, 1994). This in turn could explain where the carbon dioxide from the atmosphere became sequestered (as organic compounds) during ice ages (Section 4.3.3). However, MSA concentrations in the GRIP core do not show similar variations to those of Vostok: MSA concentrations during the last ice age were a little lower than they were during the last interglacial. This has led Legrand (1997) to suggest that climatic change promoted different responses in the sulphur cycle in the two hemispheres.

4.4 Physical characteristics

The physical characteristics of ice caps that constitute valuable tools for palaeo-environmental reconstruction are the identification of annual increments, the record of dust and its characteristics, and the presence of volcanic ash layers. All provide opportunities for correlation between cores as well as information on past environmental conditions.

4.4.1 Annual increments of ice

As snow accumulates, each annual increment comprises a dark layer, representing the summer season, and a light layer representing winter snowfall. (There are similarities here with annual varves; Section 1.2.2.) As snow and ice continue to accumulate, the layers become compressed, but annual increments can still be identified using light transmission techniques or X-ray methods. The identification of annual increments provides a means of age determination; in particular, ice-accumulation years can be established. This reflects the number of annual layers below the present surface. The identification of annual layers becomes increasingly difficult with depth because of distortion and deformation, but in many cases ice-accumulation layers have been established for part, and in many cases, most of the Holocene (Bard and Broecker, 1992). The establishment of ice-accumulation layers is also important for studies of recent environmental change, which focus on pollution (see discussion in Mannion, 1997b).

4.4.2 The dust record

Even tiny dust particles in ice can be detected by techniques based on the scattering of light. The dust record provides valuable palaeoenvironmental information; it provides an additional means of identifying annual layers (Section 4.4.1) and, if the concentration of dust particles, is determined, a record of aeolian activity. Moreover, the composition of the dust provides a means of identifying source areas. For example, Basile *et al.* (1997) have used the isotopic composition, notably of strontium and neodymium, to determine the source of dust deposited at Vostok and Dome C (see Figure 4.1 for location) during cold stages (oxygen isotope stages 2, 4 and 6). Their analyses of core material and material from potential source areas in Australia, New Zealand, southern Africa and South America shows the dust came from Patagonia. Relatively short cold periods such as the Younger Dryas *c.* 12K years BP (Zielinski and Mershon, 1997) and the Little Ice Age *c.* 600 years BP (O'Brien *et al.*, 1995), from the GISP2 and GRIP cores respectively, have also been shown to be characterised by high inputs of dust.

4.4.3 Volcanic ash

Volcanic ash, identified by its mineralogical characteristics, can contribute to palaeoenvironmental reconstruction by providing marker horizons for correlation between ice cores, marine-sediment cores, peat deposits, lake sediments and historically documented eruptions. In addition, volcanic ash layers provide evidence of volcanism prior to the historic period when records began to be kept.

The record of volcanic ash complements the record of volcanism reflected in acidity profiles (Section 4.3.6). The value of volcanic ash analysis in ice cores is illustrated by work on the GISP2 core. Ram *et al.* (1996) provide evidence, notably rhyolitic glass shards, for a volcanic eruption which is dated to 57.3K years BP and which correlates with an ash layer found in sediment cores from the Atlantic Ocean.

4.5 The ice-core story

Ice cores have provided a wealth of information on palaeoenvironmental conditions, especially over the last glacial/interglacial climatic cycle. As well as the detail of small-scale variations in, for example, acidity and heavy-metal concentrations, ice-core data provide a record that can be correlated with marine-sediment cores (Chapter 3) and continental deposits (Chapter 5) to build up a scenario of global environmental change. Much has been written about the meaning and value of ice-core records (e.g. Oeschger and Langway, 1989) and there is no doubt that the extraction of the Vostok core in particular, because of its depth, was a major watershed in palaeoenvironmental studies. Much of the detail given above refers to the Vostok or the deep Greenland cores and explanations are given in relation to the significance of individual indices.

In terms of the major changes that occurred over the last climatic cycle, the ice-core data confirm the significance of the Milankovitch (astronomical) **forcing** of climatic change (Section 2.4; Box 2.2). Evidence for environmental change with periodicities of all three of the Milankovitch components can be identified in ice-core data. These periodicities can also be found in the marine-sedimentary record. This correspondence confirms the global impact of Milankovitch forcing. Of particular significance for correlation between ice cores and marine sediments are the oxygen isotope data, which are also a proxy record of palaeotemperatures and global ice volume. Much attention has been focused on the changes that occurred as an ice age ended or began; ice-core oxygen isotope signatures indicate a temperature difference of *c.* 10°C between an ice age and an interglacial stage. Moreover, the oxygen isotope stratigraphy indicates, as it does in marine sequences, a relatively rapid shift from the glacial to the interglacial state. This is of especial interest because of the climate dynamics involved and the ecological dynamics of the resulting positive **feedback**; there may be parallels, and certainly lessons to be learnt, about the possible dynamics of current and future global warming. The specific changes are summarised in Figure 4.10, which indicates not only the magnitude but also the complexity of climatic change over this period. The global significance of these climatic changes is also highlighted by the fact that they are recorded in cores from the Huascarán glacier in Peru (Thompson *et al.*, 1995) and the Guliya ice cap on

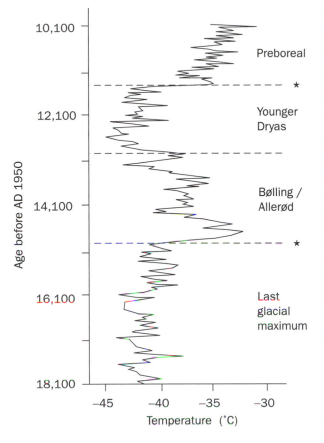

* note rapid rise in temperature

Figure 4.10 *Reconstructed temperature changes for the last c. 18K years from the GISP2 (Greenland) ice core*

Source: Based on Kapsner *et al.* (1995).

the Qinghai–Tibetan Plateau (Thompson *et al.*, 1997). The lesson from the ice cores is that abrupt climatic change is a characteristic of the past and may therefore occur in the future.

The proxy records of temperature during glacial stages are also interesting. Figures 4.2 and 4.3 show that glacial stages were far from uniform in terms of temperate regimes. The Vostok record (Figure 4.2), for example, indicates that there were at least two less climatically severe episodes, i.e. interstadials, during the 100K years of the last ice age. Lorius and Oeschger (1994) state that: 'The new Summit GRIP core indicates that such events extended over the entire ice age, with warmer interstadials beginning abruptly, perhaps within a few decades, and lasting 500–100 yr.' They also suggest that the cold glacial stages in Greenland were *c.* 12°C colder than at present, and that the relatively mild interstadials were 5°C colder than the present climate. Dust and heavy-metal records indicate that the last and penultimate ice ages were relatively arid, with aeolian erosion a common process in continental regions adjacent to the polar ice caps.

The analysis of heat-trapping gases in bubbles from ice cores has also provided vital information on natural environmental change. As Figure 4.5 illustrates, there is a high degree of correspondence between the indicators of palaeotemperatures and carbon dioxide and methane concentrations. This relationship complicates the nature of the climatic change by implicating changes in the global carbon cycle as well as Milankovitch forcing. Arguments have been advanced for changing productivity in the oceans (Section 4.3.7) as a means of altering atmospheric carbon dioxide concentrations, and for the increase and decline in the extent of global wetlands as a control on atmospheric methane. As discussed in Section 4.3.3, this is a complex issue that links Milankovitch forcing, climatic change and the global carbon cycle. The detail of this relationship has yet to be determined, but it confirms the role of heat-trapping gases in rapid climatic change. In view of the fact that human activity has increased atmospheric carbon dioxide concentrations by *c.* 25 per cent and methane concentrations by *c.* 100 per cent in the past two centuries, the lessons from the past confirm that society is right to be concerned about global warming.

4.6 Conclusions

Ice cores provide invaluable information on past environments. The data complement those from marine sediments and continental sequences and thus contribute to the development of a framework for global environmental change. Analyses of polar ice cores have made a substantial contribution to palaeoenvironmental research, but high-altitude cores from Peru and Tibet have increased confidence in ice-core data by providing corroborating data. Despite these positive aspects of ice-core data, there remain many uncertainties; techniques of analysis require refinement, as do correlations/relationships with other palaeoenvironmental records.

Most importantly, however, there remains a huge potential for ice-core research. For example, at least 1 km of ice remains to be extracted from the Vostok site; this would extend the ice-core record back in time but, inevitably, the interpretation of such a compressed record of environmental change would be difficult. In addition, there are many possibilities for ice-core work in high-altitude regions.

Summary Points

- The longest records have been obtained from Vostok in Antarctica and from near the summit of the Greenland ice cap (the GRIP and GISP2 cores).

- The longest record extends back *c.* 250K years.

- Ice cores have also been obtained from high-altitude locations in Peru and the Tibetan Plateau.

- Oxygen isotope stratigraphy provides a record of temperature and ice volume change, as well as a means of correlation with marine, loess and lacustrine oxygen isotope stratigraphies.

- Chemical indices include anions, deuterium, sodium, heavy metals, beryllium and methanesulphonate.

- Concentrations of carbon dioxide and methane in ice bubbles reflect atmospheric concentrations of these heat-trapping gases for the time period represented by the core.

- Physical indices include dust, annual layers and volcanic ash.
- The indices provide detailed information on the late **glacial/interglacial** cycle.

General further reading

Global Environmental Change. A Natural and Cultural Environmental History. A.M. Mannion. 1997. Longman, Harlow, Essex, 2nd edn.

'Greenland ice cores: frozen in time'. R.B. Alley and M.L. Bender. 1998. *Scientific American* **278**, 66–71.

Ice Age Earth. A.G. Dawson. 1992. Routledge, London.

Reconstructing Quaternary Environments. J.J. Lowe and M.J.C. Walker. 1997. Longman, Harlow, Essex, 2nd edn.

The Environmental Record in Glaciers and Ice Sheets. H. Oeschger and C.C. Langway Jr (eds). 1989. John Wiley and Sons, New York.

⬤5 The record of environmental change in continental archives

5.1 Introduction

The continental deposits that initially inspired natural scientists to speculate on past environmental change were those left by glaciers and ice sheets. Together with observations in actively glaciated regions, such deposits inspired the natural scientists of the nineteenth century to formulate the glacial theory (Section 1.2). Today, the deposits left by ice sheets, etc. still attract considerable attention, as efforts to reconstruct the dimensions and impact of the many ice advances of the last 3×10^6 years continue. Geomorphological and lithological features both have a role to play in elucidating past environmental change.

Although only high latitudes, and, to some extent, middle latitudes, as well as high altitudes, were directly affected by glacial and periglacial activities, the considerable drop in global temperatures during the ice ages had ramifications for all Earth-surface features and processes. World biomes, for example, were quite different during ice ages, in relation to both their geographical location and their species composition, compared with their characteristics during **interglacials**. Similarly, coastlines and coastal features were very different during ice ages as compared with interglacials; the incarceration of a vast volume of water caused substantial falls in sea-level and major rises ensued as ice sheets melted. What are today continental shelf regions of the ocean floor were dry land during the ice ages.

Evidence for these environmental changes derives from a variety of continental archives. On the basis of physical, chemical and biological techniques, data on environmental change are being elicited to provide a local record of change, which also contributes to the regional and global tableaux. Such archives are diverse; they include lake sediments, peats, palaeosols, loess, carbonate deposits and packrat middens. Some of these archives cover long periods of time, as is the case with a number of lake sediments and loess deposits. Others are particularly relevant to the Holocene. The use of age-determination techniques, oxygen isotope stratigraphy and palaeomagnetic stratigraphy have facilitated correlation between diverse and distant continental records and between continental records and those from ocean sediments (Chapter 3) and ice cores (Chapter 4).

5.2 Glacial and periglacial landforms and sediments

The presence of glacial and periglacial landforms provides direct evidence of past climatic regimes. The value of such deposits has been widely documented, e.g. Benn and Evans (1997), Lowe and Walker (1997) and Williams *et al.* (1998), and so reference here is brief. The presence of glacial landforms such as moraines and outwash features reflects direct glacial activity and their distribution helps to determine the maximum extent of ice sheets and glaciers. The landforms produced in a former glacial age are many and varied, as illustrated in Figure 5.1. Similarly, the presence of periglacial landforms, such as solifluction lobes and patterned ground (Figure 5.1), reflect the operation of freeze–thaw processes which are prevalent in areas close to but not directly occupied by ice sheets, etc. The sedimentary and organic characteristics of glacial (Table 5.1) and periglacial landforms also provide information on past environmental change. For example, glacial deposits are often referred to as 'till' (the landform is a moraine); this can be unstratified or stratified.

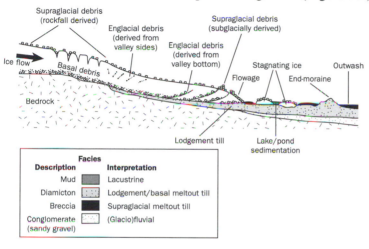

Figure 5.1 *Types of glacial deposits*

This characteristic distinguishes material deposited by glaciers (the unstratified till) from that which has been stratified or sorted by glacial meltwater. Such characteristics themselves reflect the operation of processes under specific environmental conditions. In addition, the orientation of stones within unstratified drift (also known as a glacigenic deposit: Table 5.1) and their mineralogical characteristics can facilitate the establishment of the ice-flow direction and source area of the material.

Periglacial landforms and sediments (French, 1996) reflect the operation of gelifluction processes; these involve the freeze–thaw cycles which physically break up rocks to produce angular debris and this, in turn, is transported downslope through flow and creep. Moreover, today's periglacial environments are characterised by permafrost in which ice crystals, and sometimes larger ice wedges, are formed. On permanent melting, the resulting gap becomes infilled with a type of sediment that is distinct from the surrounding material. Where a large ice mass becomes buried, and water is forced upward, a specific type of landform, known as a pingo, results. This can be recognised by a dome-shaped

Table 5.1 *Various types of glacial deposits*

....

1 Glacigenic deposits

    Supraglacial till: *Flow till* originates as material
                         saturated with water, which behaves
                         as a fluvial mass, and which is released
                         from a downwasting glacier

                         Meltout till is debris released from
                         within the glacier as ice melts above
                         the base or at the surface

    Subglacial till: *Meltout till*, as above

                         Lodgement till occurs at the base of a
                         glacier through pressure between the
                         ice and its bed

                         Deformation till is lodgement till that
                         has been affected by deformation

2 Glaciofluvial   *Ice-contact deposits* which accumulate
  deposits: in contact with the ice

                         proglacial (outwash) *deposits* which
                         accumulate near the forward margin of
                         the ice

3 Glaciolacustrine Sediments accumulate in lakes that
  deposits: are dammed by or fed by glacial
                         meltwater

                         Varves (annual layers) may form under
                         such conditions

....

structure or a rampart surrounding a depression. Such features have been recognised in Britain, notably in Wales (Ballantyne and Harris, 1994).

A major objective of research on glacial landforms, etc. is the reconstruction of the spatial and temporal variation of former glaciers and ice sheets and the conditions under which they waxed and waned. Essentially, this is the modelling of past conditions, and is exemplified by a considerable volume of work on the Laurentide ice sheet (the last extensive ice sheet in North America; Figure 5.3), and the distribution of ice in North-west Europe, including Britain, during the last glacial stage. Much of this work has been summarised by Lowe and Walker (1997). For example, the ice sheet covering Fennoscandinavia during the last glacial maximum attained a thickness of more than 2500 m, with a domed centre over the Gulf of Bothnia. Modelling of the Laurentide ice sheet also indicates that the thickness of the ice varied, possibly with centres of accumulation in the Hudson Bay area, Keewatin, Labrador and Baffin Island; ice depths of as much as 4000 m may have developed (Dawson, 1992). Inevitably, changes occurred as ice sheets developed and subsequently melted, and attempts have been made to reconstruct the dynamics of past and present ice sheets in relation to past real or likely future climatic change. For example, Siegert (1997) has examined the record of glaciation during the last glacial stage in the Barents Sea. Here, the ice sheet began to grow after *c.* 25K years BP, and started to decay *c.* 16.5K years BP; these age estimates reflect the later accumulation and earlier demise of this ice sheet in comparison to others such as that of Fennoscandinavia. Why this variation should have occurred is open to speculation, but it illustrates the non-uniform response of ice sheets. This may well occur in the future, especially in relation to the Greenland and Antarctic ice sheets, should global warming intensify.

Another important objective of research on glacial and periglacial landforms and sediments is the determination of ice limits during past ice advances. Such reconstructions are difficult because the erosional power of one ice advance can readily eradicate the moraines, etc. of earlier ice advances. This fragmentation of the terrestrial record makes the interpretation of environmental change particularly difficult (Section 1.2.1). This is especially so for glaciations prior to the last one. The glaciation history of various parts of the world has been reviewed by Dawson (1992) and Williams *et al.* (1998). Figure 5.2 illustrates the extent of the ice during the last three ice ages in Britain. It shows that the oldest of these, the Anglian (see Figure 2.4 for correlations with Europe and North America), was the most extensive, extending as far south as a line connecting London and Bristol. The area south of the limits of the last three major ice advances experienced periglaciation (see above). Figure 5.3 illustrates the extent of ice cover during the last glacial stage in North America, where the ice extended further south than the Great Lakes. Figure 5.4 shows the global extent of ice during the last glacial stage, as compared with today. It gives some idea of the vast volume of ice that was present, especially in the Northern Hemisphere. Moreover, the ice sheets were more than 1 km deep in places. With so much water incarcerated in ice, sea-levels were much lower than they are today. The world was indeed a different place!

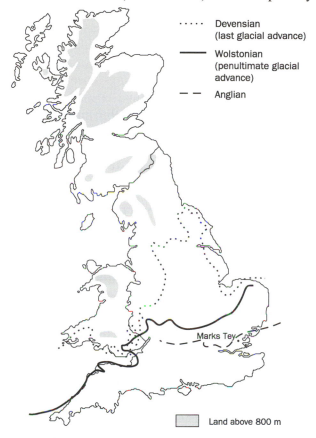

····· Devensian
(last glacial advance)

——— Wolstonian
(penultimate glacial advance)

— — Anglian

Marks Tey

☐ Land above 800 m

Figure 5.2 *The maximum extent of ice cover in Britain during the last three major glaciations*
Source: Based on maps in Jones and Keen (1993).

5.3 Lake sediments

As soon as a topographic hollow becomes filled with water to form a lake, sediment begins to accumulate. This sediment comprises soils and unconsolidated materials eroded from the catchment, and, as it settles out of the water it entrains additional material, such as algae, insects, molluscs and

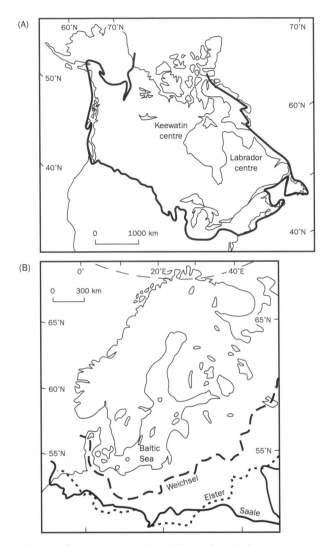

the remains of the vegetation that grew in the lake. It also incorporates particles that entered the lake from the atmosphere, e.g. pollen grains produced by the vegetation of the catchment. There are established techniques for analysing the pollen content of lake sediments and peats, as detailed in Box 5.1. Lake sediments thus provide a chronicle of the environmental change since the deposition began. The plants and animals that live within and around the lake are influenced by the prevailing climate, so their remains in the lake sediment also provide information on past climates.

In areas of the world that were not directly glaciated (Figure 5.4), lake sediments had been accumulating throughout the Quaternary period, and in some cases they had been accumulating for the last 3×10^6 years, e.g. at Lake Biwa in Japan. Such sequences are particularly valuable because they provide a long and unbroken record of environmental change in continental contexts. This record can be correlated with ocean sediments (Chapter 3), and thus contributes to the understanding of global environmental change. In those areas of the world that have experienced glaciation, the accumulation of lake sediments began

Figure 5.3 *(A) The major ice limits during the last (Wisconsin) ice advance in North America (based on Aber, 1991). (B) The southern margins of the last three major ice advances in Europe (based on Donner 1995).*

only after the end of the last glacial stage. Such sediments provide a chronicle of environmental change for the last 14K years at the most. They provide information on natural environmental change during the first half of the present **interglacial** (the Holocene), as well as information on human impact in the later part. In addition, lake sediments deposited during earlier interglacials, and which escaped the erosive action of subsequent glaciations, provide insights into environmental change.

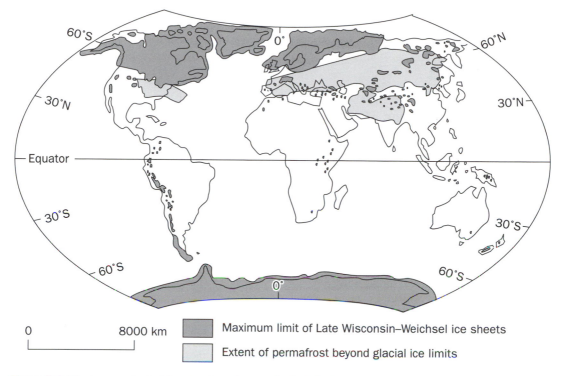

Figure 5.4 *Maximum extent of ice sheets and permafrost at the last glacial maximum*
Source: From Williams *et al.* (1993).

5.3.1 Long lake sequences

Most long lake sequences occur in middle and low latitudes, as shown in Figure 5.5. One of the longest sequences comes from Funza, near Bogotá in the Eastern Cordillera (Andes) of Colombia; it covers *c.* 3×10^6 years. Some 55 pollen assemblage zones have been identified, which relate to at least 27 climate cycles (i.e. glacial–interglacial cycles). The pollen assemblages have been used to determine changes in the height of the tree-line: during the warm stages the forest reached an altitude of *c.* 3000 m with perennial snow above 4500 m, but during cold stages the tree-line was *c.* 1000 m lower at an altitude of 2000 m with perennial snow above 3500 m (Hooghiemstra *et al.*, 1993). Between the tree-line and snow-line, various types of grassland predominated. These oscillations are reflected in Figure 5.6, which is concerned with environmental change during the last 800K years, as reflected in several long sequences in various parts of the world. The Funza data, and those from other sequences, have been correlated with marine oxygen isotope stratigraphy. In terms of climatic change, the variations in the Funza tree-line indicate a temperature difference of *c.* 9°C during a glacial–interglacial cycle. Changes of a similar order of magnitude are implied by ice-core data from Huascarán in Andean Peru (Thompson *et al.*, 1995) and by pollen data from Lake Biwa, Japan (Fuji and Horowitz, 1989), as shown in

Box 5.1

The principles of pollen analysis

1 Pollen analysis was developed as an analytical technique in the early 1900s, and since then it has become one of the most widely applied techniques in palaeoenvironmental research.

2 Pollen is produced by higher plants (e.g. trees, shrubs, herbs), and spores are produced by lower plants (e.g. mosses and ferns). Both pollen and spores are emitted into the atmosphere and eventually settle out on land or water. Most are oxidised, i.e. degraded to their basic components. Others become entrained in peat bogs and lake sediments. In such waterlogged environments, pollen grains and spores are well preserved. As the peat or lake sediment accumulates, the layers which contain the pollen and spores provide an archive of environmental change.

3 Peat and lake sediments can be sampled using coring apparatus. The length of core represents a temporal sequence, with the lower layers representing the period when accumulation began and the top layers representing the time when accumulation ceased, or the present day.

4 Using sub-samples of c. 1 cm^3 from various layers in the core and employing a variety of chemical processes to remove the siliceous and organic materials it is possible to isolate the pollen and spores (see below). These can be identified under the microscope to genus level by comparing the pollen/spores with keys or a type collection. Based on counts of 300+ grains, a percentage pollen diagram can be produced (see Figure 5.8). This reflects changes in the plant community for the length of time represented by the core. It is then possible to consider what the stimuli were for such changes.

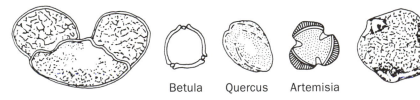

Betula   Quercus   Artemisia

Pinus (Pine)   Tilla (Lime)

5 Radiocarbon age determination (see Box 5.2) can be applied to ascertain the ages of specific layers either by using seeds or plant remains directly, or by using the organic content of the sediment/peat matrix.

See Moore *et al.* (1991) for details.

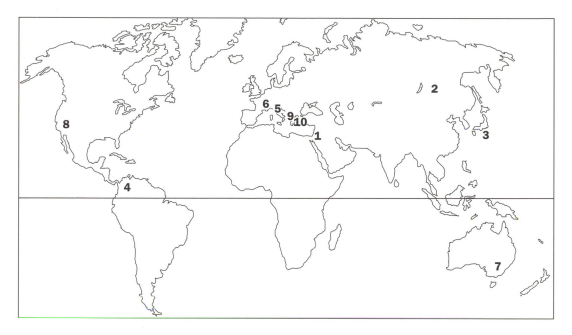

1. Lake Hula, Israel
2. Lake Baikal, Russian Federation
3. Lake Biwa, Japan
4. Funza, near Bogotá, Colombia
5. Valle di Castiglioni near Rome, Italy
6. Praclaux Maar Lake, Massif Central, France
7. Lake George, New South Wales, Australia
8. Owens Lake, California, USA
9. Tenaghi–Philippon, Greece
10. Ioannina, Greece

Figure 5.5 *Locations of long lake sequences referred to in the text*

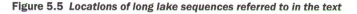

Figure 5.6. The Hula Basin (Israel) record reflects the predominance of relatively arid conditions during interglacial stages. This contrasts with evidence from high latitudes for an increase in precipitation during interglacials. Furthermore, it contrasts with the pollen record from Lake George, New South Wales, Australia (Singh and Geissler, 1985), which is also shown in Figure 5.6. The inference from the Lake George data is that **sclerophyll woodland** dominated interglacials, in contrast to the open herbaceous communities that dominated the glacial stages.

Additional long pollen sequences reflect corresponding and synchronous changes in vegetation communities. For example, the Valle di Castiglioni near Rome provides a record for the last 250K years (Follieri *et al.*, 1988), while the Praclaux **maar lake** (a lake in the crater of an extinct volcano) on the Devès Plateau of the Massif Central, France, contains a record for the last 400K years (Beaulieu and Reille, 1992). There are two long sequences from Greece: Tenaghi–Philippon, which extends back 1×10^6 years (Mommersteeg *et al.*, 1995), and Ioannina with a 423K year record (Tzedakis, 1994). The latter represents **oxygen isotope stages** 1–11, i.e. five interglacials and four cold stages; during the warm stages, forests of oak, elm, hornbeam, fir, beech and hop hornbeam predominated, whilst during the cold stages forest–steppe and desert–steppe communities occurred.

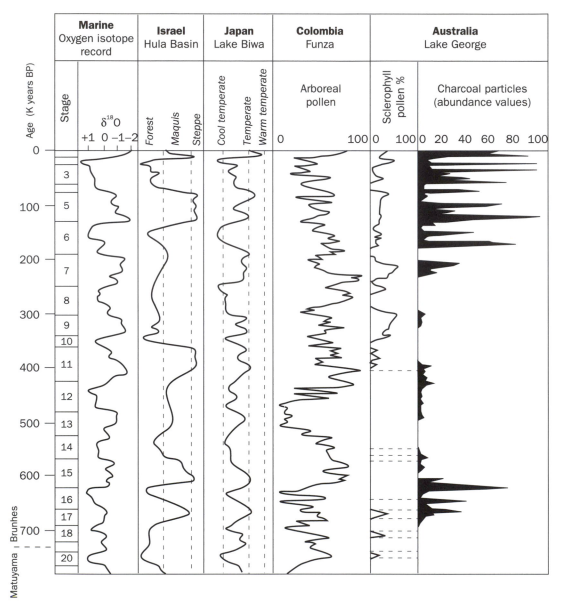

Figure 5.6 *Correlations between four long lake sequences and the marine oxygen isotope record*

Source: From Williams *et al.* (1993).

Several long lake sequences have been investigated in the USA. For example, the sediments of Owen's Lake playa, Inyo County, California, have been examined by Bradbury (1997), a simplified version of whose results are summarised in Table 5.2. In this case, diatom (siliceous algal microfossils; Section 3.2.3) assemblages, as well as pollen assemblages and a variety of sediment characteristics, were used to reconstruct the lake's characteristics for the last 200K years. As in the case of the Hula Basin of Israel (see above and Figure 5.6), the interglacial stages were

Table 5.2 *A summary of environmental change over the last 200K years, from Owen's Lake, California*

Approx age (K years BP)		Diatom assemblages	Dominant pollen	Sediment characteristics
	Present interglacial (the Holocene)	Saline diatom species		
10				
	Last glacial stage	Freshwater planktonic species	Juniper	Abundant plagioclase
85				
	Last interglacial	Saline diatom species		
100	Penultimate glacial stage	Freshwater planktonic species	Juniper	Abundant plagioclase
180				
200				

Source: Based on Bradbury (1997).

characterised by arid and saline conditions. In the Owens Lake region, juniper communities dominated the catchment during the glacial stages when freshwater conditions prevailed. In another long sequence from Lake Baikal, Siberia, **biogenic** silica records have been shown to correlate with oxygen isotope stages (Colman *et al.*, 1995); more recently these records have been correlated with those from Lake Biwa (Xiao *et al.*, 1997). In particular, low influxes of biogenic silica occurred during **stadial** periods, while high influxes characterised interglacials and **interstadials**. Xiao *et al.* have recorded five periods of high biogenic silica influx, which correlate with oxygen isotope stages 5e, 5c, 5a, 3 and 1; conversely, periods of low biogenic silica influx occurred during oxygen isotope stages 6, 5d, 5b, 4 and 2. Overall, the Lake Biwa and Lake Baikal data indicate that diatom productivity, for which biogenic silica is a proxy measure, was greatest during periods that were warm and wet, notably the interglacial and interstadial periods.

5.3.2 Interglacial lake sediments

The long lake sequences referred to in Section 5.3.1 contain records of past interglacials, which thus provide evidence of natural environmental change during these stages. For example, the Funza sequence from the High Plain of Bogotá (Hooghiemstra, 1989) shows that during the last interglacial stage, oak (*Quercus*) forests were dominant in the catchment, with alder (*Alnus*) carr in flat areas around the lake. The record also shows that alder migrated into the region from North America, *c.* 2.7×10^6 years ago as a consequence of the closure of the Panama Isthmus, with oak arriving *c.* 1×10^6 years ago. Thereafter, these species dominated interglacial forests *c.* 2500 m above sea-level.

In the high-latitude and middle–high latitude zones that were subject to direct glaciation, the record of past interglacials is represented in lake deposits that were just beyond the limits of subsequent glacial deposits. In Britain, for example, lake sediments (and fluvial deposits) have been found at numerous locations in East Anglia and south-east England (Jones and Keen, 1993). A classic sequence for the Hoxnian interglacial (marine oxygen isotope stage 9; Figure 2.4) has been examined from Marks Tey, near Colchester in Essex (see Figure 5.2 for the location); these sediments represent a complete interglacial stage, and its pollen assemblages, along with those from nearby but incomplete sequences, provide information on vegetation change. The characteristics of the interglacial vegetation communities are summarised in Table 5.3. Initially, an open habitat community is replaced by birch (*Betula*) forest with pine (*Pinus*); this in turn is replaced by deciduous forest and finally heathland and grassland.

Table 5.3 *The vegetation changes in East Anglia, UK, during the Hoxnian interglacial stage*

Approx. age (K years BP)				Oxygen isotope stages
	PENULTIMATE (WOLSTONIAN?) GLACIATION			8
300 —				
	Ho IV	b	Reduction is alder and fir: in-spread of birch as grassland expanded and heath declined	
		a	Fir remained important, with in-spread of pine with crowberry heath and grassland	
	Ho III	b	Predominance of alder as hornbeam and hazel decline and fir increases	
		a	Spread of hornbeam, decline of oak and elm	9
	Ho II	c	Elm, hazel and yew were significant *	
		b	Alder and hazel became important	
		a	Oak predominated	
	Ho I		Birch forest with pine	
330 —				
	ANGLIAN	GLACIATION		10

* At several sites an episode of deforestation is recorded, possibly due to burning.

Source: Based on West (1980).

In Central Europe, lacustrine deposits from Belchatow in Poland have been investigated by Krzyskowski *et al.* (1996) in order to determine the characteristics of environmental change during the Ferdynandovian interglacial (marine oxygen isotope stage 11; Figure 2.4). The sediments comprise diatom-rich clays and **gyttja** (an organic mud), and analyses of pollen and carbonates show that as the climate ameliorated at the opening of the interglacial, the biological activity increased in the basin. Moreover, temperature reconstructions based on isotope analyses indicate a temperature change of *c.* 7.5°C from the beginning of the interglacial to the initial appearance of deciduous trees.

Pollen analysis (Box 5.1) has been widely used elsewhere in Europe to reconstruct interglacial environments. For example, Figure 5.7 is a generalised reconstruction of vegetation communities in north-west Europe during the last (Eemian: marine oxygen isotope stage 5e; Fig. 2.4) interglacial stage. During this period, 115K to 130K years BP, temperatures were probably warmer than today by *c*. 2°C, and sea-levels were higher than at present; the forest extended further north than at present and Fennoscandinavia was an island (Andersen and Borns, 1994).

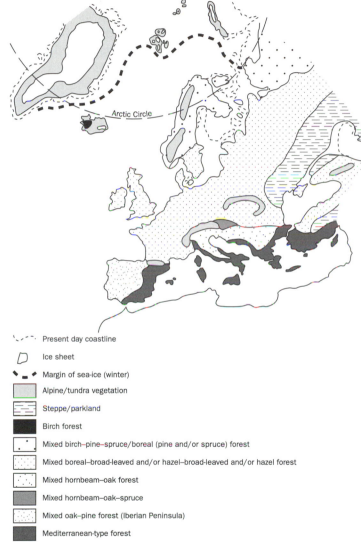

Arctic Circle

'\.-⁻˙ Present day coastline

🗋 Ice sheet

◆_▬◆ Margin of sea-ice (winter)

▨ Alpine/tundra vegetation

▤ Steppe/parkland

■ Birch forest

⬚ Mixed birch–pine–spruce/boreal (pine and/or spruce) forest

⬚ Mixed boreal–broad-leaved and/or hazel–broad-leaved and/or hazel forest

⬚ Mixed hornbeam–oak forest

▨ Mixed hornbeam–oak–spruce

⬚ Mixed oak–pine forest (Iberian Peninsula)

■ Mediterranean-type forest

Figure 5.7 *Land-cover types during the Eemian interglacial*

Source: Based on Andersen and Borns (1994).

5.3.3 Postglacial lake sequences

The majority of lake sequences, particularly in active lakes in the temperate zone, relate to the period after the last ice age. The first *c*. 4K years after the last ice age is known as the late glacial, and is highly complex in terms of the environmental changes that occurred (this will be discussed further in Chapters 7, 8 and 9); the Holocene, the present interglacial, began at *c*. 10K years BP. Together these periods comprise the postglacial period. In areas that were glaciated, erosion by ice gouged out hollows that subsequently became lake basins; in the upland zones the former lake basins were themselves often centres of ice accumulation and after the ice melted they filled with water. In addition, lakes sometimes formed at the edge of ice sheets (Section 5.2), though not always permanently, and pools and ponds formed in topographic hollows and valleys that had been subject to periglaciation. In these situations, sediments began to accumulate as the postglacial period opened. The physical, chemical and biological characteristics of such sedimentary archives provide a wealth of information on postglacial environments at a wide variety of altitudes and latitudes.

Indeed, much of the seminal research on palaeoenvironments in the 1920–1950 period focused on lacustrine environments. For example, much of Godwin's work in East Anglia, England, involved the analysis of pollen from the region's numerous meres. A classic pollen diagram from Hockham Mere is illustrated in Figure 5.8. This shows that during the early stages of the postglacial period, open-habitat communities prevailed, and that the pioneer tree species were birch (*Betula*) and pine (*Pinus*). Eventually a mixed-deciduous forest developed, dominated by elm (*Ulmus*), alder (*Alnus*), lime (*Tilia*) and oak (*Quercus*). This occurred by *c.* 5.5K years BP, after which time human activity greatly influenced vegetation change.

Another example of the detection of environmental change from lake sediments is that of Moore Lake, Alberta, Canada. This study involved a multidisciplinary approach: using pollen, diatoms, **chrysophyte** remains (planktonic algae with siliceous scales) and sedimentary pigments, Hickman and Schweger (1996) have reconstructed the postglacial history of the lake and its catchment. A summary of their results is given in Table 5.4. As in the case of Hockham Mere, the pollen assemblages reflect the vegetation succession that occurred as the ice retreated; by 6.2K years BP, birch (*Betula*) and spruce (*Picea*) forest had developed in the catchment, and a diatom flora indicative of saline conditions had developed

Table 5.4 *Environmental changes of the last 12K years in Moore Lake, Alberta, Canada*

Approx. age (K years BP)	Vegetation	Diatom flora	Diatom productivity
0	Boreal birch–pine–poplar forest		
2		Freshwater Planktonic	High
4	Boreal birch–spruce–alder forest	Oscillating planktonic, saline and freshwater spp.	Mixed with peaks and troughs
6		Saline spp.	
	Parkland/ grassland	Most saline spp.	Very low
8		Mixed saline and freshwater spp.	
	Open birch–spruce forest	Most saline spp.	Very low
10			
	Open spruce–birch forest	Freshwater planktonic	High
12			
	Pioneer	Pioneer benthic and epipelic spp.	Low

Source: Based on Hickman and Schweger (1996).

in the lake. This saline episode, along with others between 5.8K years and 4.0K years BP, reflects the prevalence of relatively arid conditions.

The value of lake sediments for providing a record of both natural and anthropogenic environmental change is also exemplified by the recent review by Berglund *et al.* (1996a) on the palaeoecology of Europe during the last 15K years. This involves regional syntheses of lake and mire records, as well as the establishment of regional-scale and continental-scale correlations. Such syntheses are particularly valuable for determining the dynamics of ecosystems, including the location of **refugia** and migration rates, during the postglacial period. Moreover, much effort has been invested in the reconstruction of lake pH histories using diatom remains (Figure 3.4), e.g. Battarbee (1994), though this relates to culturally induced rather than naturally induced environmental change (see review in Mannion, 1997b).

5.4 Peats

Peats are another archive of environmental change. Their sequential accumulation over time preserves a record of peatland vegetation and, as in the case of lakes, they absorb particulate matter from the atmosphere. This includes pollen grains, so peatlands contain a record of regional vegetation change as well as local vegetation change. In the high and middle–high latitudes peat accumulation began during the postglacial period in areas previously glaciated or periglaciated. However, in low latitudes, the record of environmental change in peatlands extends further back in time, though most peatlands occur only in high latitudes where precipitation is sufficient to maintain waterlogging.

In Britain in the 1920s, some of the earliest pollen diagrams were constructed on the basis of research on Scottish highland and island peats. In addition, pollen analysis of Fenland peats in East Anglia confirms the regional record of vegetation change derived from lake sediments such as those of Hockham Mere (Section 5.3.3). Much of this work has been synthesised by Godwin (1975), Jones and Keen (1993) and Berglund *et al.* (1996a). Pollen assemblage data from peats have been combined with those from lake sediments to reconstruct the characteristics of regional forest types in Britain and Ireland *c.* 5.5K year BP. The results of this synthesis are presented in Figure 5.9. This particular point in the postglacial period is important because it represents the optimum period of forest development prior to substantial modification by human activity. Moreover, Figure 5.9 reflects the considerable regional differences in forest composition that existed at this time in response to soil characteristics as well as altitudinal and latitudinal climatic gradients.

Beyond Britain and Ireland, palaeoenvironmental research on peatlands has been extensive, using established techniques such as pollen and plant macrofossil

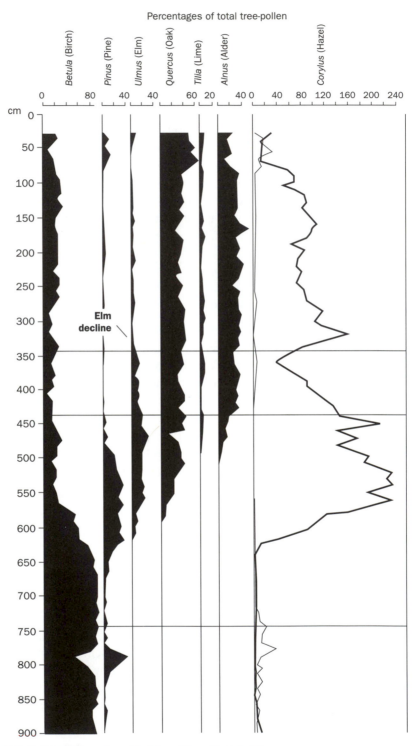

Figure 5.8 *Pollen diagram from Hockham Mere, East Anglia*

Source: Based on Godwin (1975).

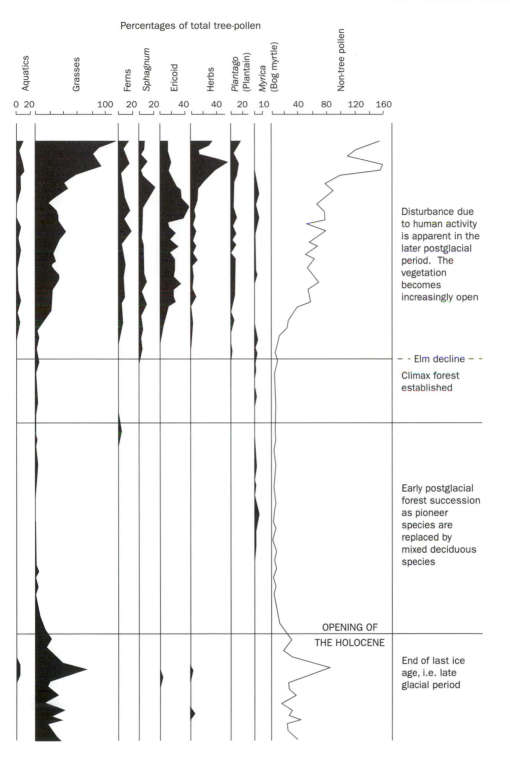

Percentages of total tree-pollen

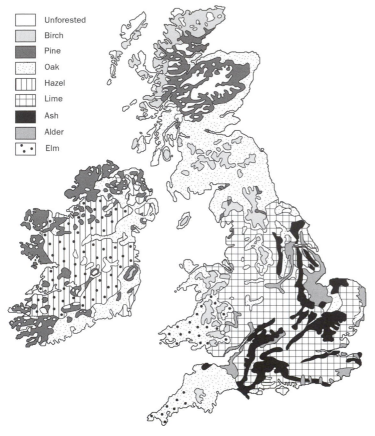

Unforested
Birch
Pine
Oak
Hazel
Lime
Ash
Alder
Elm

Figure 5.9 *The major forest types c. 5.5K years BP*
Source: Based on Bennett (1989) and Rackham (1986).

assemblages, e.g. parts of the African and South American highlands, as well as parts of upland New Zealand and Australia (notably Tasmania and New South Wales). An example of this type of research from the tropics is that on the Rusaka peat bog at 2070 m altitude in Burundi, central Africa. Here analysis of pollen and organic matter (Bonnefille *et al.*, 1995) indicate that an increase in forest cover and carbon storage, reflecting increased carbon fixation through photosynthesis, began at *c*. 12K years BP. Tree cover declined between 10.6K years BP and 10.0K years BP, a period equivalent to the Younger Dryas of Europe (Sections 4.4.2 and 8.2) when climate deteriorated.

Thereafter, forest became re-established. The evidence from Burundi indicates enhanced aridity and lends support to the suggestion that the Younger Dryas was a global event (see review in Mannion, 1997b).

5.5 Palaeosols

Palaeosols are ancient soils formed during former times on old land surfaces, and which become buried by later sediments such as **glacial** or periglacial deposits, peats or **loess**. Palaeosols thus represent past periods in time and provide the means for environmental reconstruction through pollen analysis, soil micro-morphology and organic matter content. All reflect the past conditions under which the soil developed. Palaeosols also provide stratigraphic marker horizons, rather like the volcanic ash horizons that occur in ice cores (Section 4.4.3) and in peats. Moreover, the presence of organic matter may facilitate radiocarbon age determination. The latter is described in Box 5.2 and is a particularly important

Box 5.2

The principles of radiocarbon age determination

1 Radiocarbon age determination was established by Willard Libby, an American physicist, in the 1940s and was a major breakthrough in the establishment of absolute age determination. It remains one of the most widely used techniques today, but is only applicable to organic materials.

2 Carbon exists in several isotopic forms: ^{12}C, ^{13}C and ^{14}C. Only ^{14}C is radioactive. It is formed in the upper atmosphere as cosmic radiation bombards the nuclei of ^{14}N:

$$^{14}N \xrightarrow[\text{Emission of beta particles as decay occurs}]{\text{Cosmic ray bombardment}} {}^{14}C$$

3 In the environment, ^{14}C behaves the same way as ^{12}C. Thus it oxidises to carbon dioxide in the atmosphere, and thereafter enters the actively circulating component of the global carbon cycle. It is absorbed by living organisms: plants absorb it through photosynthesis, and animals absorb it as it is passed along food chains/webs.

4 When organisms die, they no longer freely exchange ^{14}C. Consequently, the amount of ^{14}C begins to decline as radioactive decay continues and ^{14}C is not replaced.

5 The half-life of ^{14}C represents the rate at which it decays. It is the time taken for decay of half the original ^{14}C present. Libby calculated that the half-life was 5568 ± 30 years; this has since been corrected to 5730 ± 40 years, though for the sake of consistency with early dates the Libby half-life is still used.

6 The age of a sample recovered from the sediments is determined by comparing the remaining ^{14}C content with that of a modern sample: the time elapsed since death can then be calculated.

7 The measurement of ^{14}C activity can be achieved in two ways. Conventional ^{14}C age determination involves the counting of beta particle emissions. These are emitted by the ^{14}C and are measured when ^{14}C is released from the sample as a gas, e.g. $^{14}CO_2$ or $^{14}CH_4$. Accelerator-mass spectrometry (AMS) age determination involves the direct measurement of ^{14}C atoms by passing charged particles through a magnetic field at very high speeds.

8 There are many problems associated with ^{14}C age determination, including the possibility of contamination by modern material and also the counting errors associated with the techniques.

See Lowe and Walker (1997) for further details.

means of determining the age of organic materials formed during the past c. 50K years. In some situations, entire sequences of palaeosols may be identified, which are associated with features of the topography. Such a sequence is referred to as a 'palaeo-catena', i.e. a sequence of buried soil profiles that have developed over a pre-existing topography.

Boardman (1996) has presented numerous examples of palaeosols from a variety of locations. For example, he refers to palaeosols in Britain that separate glacial deposits, i.e. they represent periods of weathering of glacial moraines prior to renewed moraine deposition. One such example is Valley Farm soil in south-east Suffolk. This is enriched with clay and is red in colour, i.e. characteristics that typify temperate or even Mediterranean soils and which reflect its formation under climatic conditions that may have been warmer than today, possibly during the Cromerian interglacial. This palaeosol is sometimes overlain by another palaeosol known as the Barkham soil. This is quite different to the Valley Farm soil insofar as it contains wind-blown (aeolian) sediment and evidence of periglacial processes. It is likely to have formed during the Anglian glaciation, which followed the Cromerian interglacial (Figure 2.4). Such palaeosols are thus important stratigraphic marker horizons, as they are in other regions that were once glaciated. In addition, palaeosols can provide information on past conditions prevalent in extraglacial regions. In southern Anatolia, for example, Atalay (1996) has equated the formation of red sediments and soils with interglacials, at which time these soils were subject to the leaching of clay and carbonate. Moreover, palaeosols in loess sequences in China and Central Asia have provided valuable palaeoenvironmental information, as is discussed in Section 5.6 below.

5.6 Loess deposits

Loess is a fine-grained silt, a homogeneous and often yellow deposit which originated as wind-blown material from source areas at a distance from the area of deposition. Loess deposits cover c. 10 per cent of the Earth's surface; the most extensive deposits are in the Loess Plateau of China, Central Asia, Central Europe, Central North America and the grasslands of South America, with minor occurrences elsewhere, e.g. in New Zealand. In recent years, these extensive deposits have been the subject of intensive investigation, especially in China, and it is now possible to effect correlations between these deposits and marine oxygen isotope stratigraphy (Chapter 3).

The loess of China's Loess Plateau originated as dust blown from the Tibetan Plateau during glacial stages when arid conditions prevailed. According to Sun *et al.* (1998), aeolian sediments began to accumulate in the region as early as 7.2×10^6 years ago. At a site in the central region of the Loess Plateau, 126 m of Tertiary red clay were deposited, followed by 162.5 m of Quaternary sediments comprising

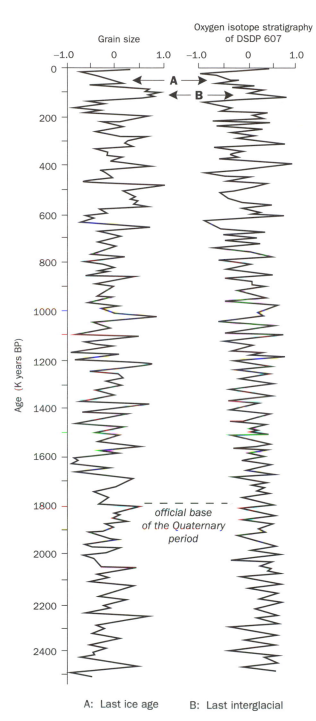

Grain size

Oxygen isotope stratigraphy
of DSDP 607

A: Last ice age   B: Last interglacial

Figure 5.10 *The Baoji loess (200 km west of Xian) grain-size record in relation to the oxygen isotope record of North Atlantic sediment core DSDP 607*

Source: Based on Ding *et al.* (1994).

loess with intercalated palaeosols. The latter were produced when, during **interglacial** stages, weathering occurred and later the soils were buried as loess deposition recurred. Such soils thus became palaeosols (Section 5.5). In parts of the Loess Plateau, the loess can be as much as 280 m deep, and it represents the last 2.6×10^6 years of deposition. Consequently, the loess is an important palaeoenvironmental archive. It has been investigated using a variety of techniques, including magnetic susceptibility measurements, particle-size analysis and the soil micromorphology of palaeosols.

Figure 5.10 illustrates the grain-size record from Baoji, 200 km west of X'ian (the city adjacent to the famous archaeological site containing the terracotta warriors) in relation to the oxygen isotope record from the North Atlantic Ocean. This shows that there is a high degree of correspondence between the curves for the last 2.5×10^6 years. The implication is that coarse loess, i.e. that characterised by a large grain size, is deposited in glacial/cold stages, having been transported by intense winter monsoons. During interglacials/warm stages the monsoon intensity was much reduced in extent and its deposited sediment is all fine grained. This relationship has also been demonstrated for other loess sequences, e.g. the Luochuan section *c.* 200 km north of X'ian (Vandenberghe *et al.*, 1997). Moreover, a similar relationship is apparent between the magnetic susceptibility of loess sequences and the marine oxygen isotope record. As Bloemendal *et al.* (1995) have pointed out, the magnetic susceptibility of loess is much higher in sediments deposited during interglacial stages than in sediments of glacial stages. They highlight the positive correspondence between the two records for profiles from the central Loess Plateau for the last 1.5×10^6 years and emphasise the presence of 100K year periodicities in the record after 1×10^6 years BP. This, in common with marine sedimentary (Chapter 3) and ice-core (Chapter 4) records, highlights the operation of the Milankovitch orbital forcing cycles (Section 2.4; Box 2.2).

The palaeosols preserved in loess sequences are not only stratigraphic marker horizons but also provide information on environmental changes during interglacial stages. For example, the various loess components indicate increased humidity when compared with loess of glacial age. In addition, the palaeosols reflect the operation of soil-forming processes, including the differentiation of soil horizons which indicates a degree of interglacial climatic stability. The occasional presence of coarse-grained horizons in palaeosols indicates periods of aridity in otherwise humid stages, while variations in magnetic stratigraphy may reflect the occurrence of natural fires. The significance of loess sequences in China (and elsewhere) is clearly growing as additional techniques are applied and new sequences are investigated.

5.7 Carbonate deposits

There are many different types of carbonate deposits that occur in regions with carbonate-rich bedrock such as chalk and limestone. For example, calcium carbonate is deposited in caves to produce a number of features, including stalagmites and stalactites. Collectively, these and other carbonate deposits in caves are referred to as **speleothems**. Sometimes the calcium carbonate may be deposited as calcite, which is its crystalline form. **Travertine** may also be found in caves. This is a light-coloured porous concretion similar to calcareous **tufa**, and which may occur as a coating on bedrock above ground where water emerges from underground. Tufa is formed as calcium carbonate is deposited around a spring or in a lake. Such deposits are valuable palaeoenvironmental indicators because they form incrementally, and in some cases the increments may be annual, rather like varves (Section 1.2.2).

Moreover, the formation of speleothems is influenced by climate. For example, during glacial stages, speleothems cease to form in regions experiencing direct glaciation because the cave systems become flooded; during interglacials their growth recommences. Thus in these regions speleothems reflect the record of Quaternary environmental change. Where the speleothem is formed under conditions that exclude contact with the atmosphere, as might be the case in deep caves, the calcium carbonate reflects the isotopic composition of the water from which it was precipitated. As a result the oxygen isotope signature can be used as an indicator of climate (Section 3.2.1; Box 3.1), as can the oxygen isotope record of calcareous lake sediments (the tufas referred to above). The ages of these calcareous deposits may also be determined using uranium-series dating (Box 5.3).

One example of a speleothem record of environmental change is that of Williams (1996), who reports on a study from Aurora Cave, near Lake Te Anau in New Zealand's fjordland. Here there are fluvioglacial sediments alternating with speleothems. Uranium-series age estimation indicates that the sediments represent the last 230K years, during which time there were seven glacial advances. Of these, five can be attributed to the last (Otiran) glaciation. Lauritzen (1995) has undertaken a similar study in northern Norway where the speleothems investigated reflect environmental change during the last 150K years. The oxygen isotope signature covers the last interglacial, the last glacial period and the Holocene. It correlates well with the GRIP (Greenland) ice-core record (Section 4.3.1), and lends support to the possibility of an unstable climate during the last interglacial.

Carbonate deposits in low latitudes can also provide valuable palaeo-environmental insights. This has been demonstrated by Crombie *et al.* (1997), who have used travertine deposits in the Kurkur Oasis of the Western Desert of Egypt to determine the environmental history of the region for the last *c*. 260K years. On the basis of uranium-series age estimation (Box 5.3) and oxygen isotope data (Box 3.1), Crombie *et al.* found that travertines were deposited around springs or within ancient lakes during the last three interglacial stages when rainfall was considerably higher than during the intervening glacial stages.

5.8 Packrat middens

Packrat middens comprise accumulations of now-fossilised plant material that were deposited by small rodents (*Neotoma* spp.) which foraged for plant food within a radius of *c*. 30–50 m from their dens. The latter were often caves and rock shelters within which the debris was deposited and cemented on to the rock surface through the solidification of the animals' urine. Such deposits were originally discovered in the Grand Canyon, Arizona, USA in the early 1980s (e.g. Cole, 1982), and since then this type of research has been extended into other parts of the arid South-west USA and arid regions elsewhere, such as in Mexico. Packrat middens provide valuable palaeoenvironmental information on several counts: they were deposited over the last *c*. 40K years and thus reflect conditions during part of a **glacial/interglacial** cycle, including the transition period; fossil plant remains, which can be identified to species level, contained within the middens, represent past vegetation communities, and if data from middens at similar elevations but different times are compared they provide information on

Box 5.3

The principles of uranium-series age-estimation methods

1 The development of uranium-series age determination is relatively recent, i.e. since the 1980s. It is based on ^{238}U, ^{235}U and ^{232}Th, which all decay to stable lead isotopes.

2 Uranium-series age determination can be used for a variety of materials, including speleothems, carbonate deposits such as travertines, tufa and calcite, as well as for molluscs, bone, shell and peat. Age determination can be undertaken for materials up to 1×10^6 years old.

3 The principle of uranium-series age determination is similar to that of radiocarbon (Box 5.2), notably that radioactive isotopes decay to a stable form at a constant rate known as the half-life. The decay series (simplified) for ^{238}U, ^{235}U and ^{232}Th are given below.

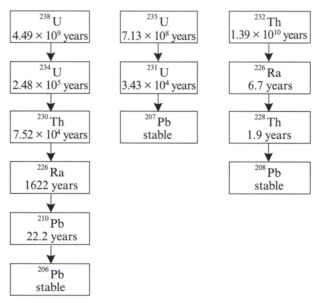

4 As daughter isotopes are formed and become separated from the parent in sedimentary systems disequilibrium occurs. Restoration of the equilibrium takes place, as time elapses, and this provides the basis for age estimation. This is also reliant on the half-life of the relevant nuclide.

See Ivanovich and Harmon (1995) for further details.

the migration of species into a region and also the altitudinal shifts of vegetation types; the plant remains contained within the middens can be dated directly by radiocarbon age estimation (Box 5.2).

Cole's original work (Cole, 1982; 1985) showed that over the last 24K years the vegetation communities changed markedly. As well as reconstructing the past patterns of vegetation distribution, Cole was able to demonstrate that altitudinal variations in past communities were not identical, in terms of species composition, to today's latitudinal distributions of vegetation. Although altitudinal shifts occurred, as climatic change occurred, species responded in an individualistic manner (Section 1.2.3). A further implication is that modern vegetation communities do not necessarily provide reliable analogues for past vegetation communities. In terms of species migration, data from packrat middens indicate that Pinyon pine (*Pinus monophyla*), which is characteristic of middle elevations in the White Mountains of Nevada and California today, entered the region at *c*. 9K years BP (Jennings, 1995). It became increasingly significant at the expense of the Utah juniper (*Juniperus osteosperma*) as the Holocene proceeded, possibly because of the establishment of dry conditions during the summer instead of an even, seasonal distribution of precipitation.

5.9 Conclusions

There is a considerable range of continental archives that contain a record of environmental change. Some of these, notably lacustrine and **loess** sequences, provide a record of the last 3×10^6 years and are particularly valuable because their records can be correlated with ice-core and ocean-sediment records. Consequently, they provide vital links between the ocean, polar and continental records of environmental change, which are essential to the formulation of a framework for global environmental change. Their provision of a long and unbroken record contrasts with other continental archives, which are often difficult to relate to a regional or global context. Glacial and periglacial deposits, for example, and especially those that belong to glacial stages earlier than the last, may be difficult to interpret in relation to their temporal distribution; incorrect interpretations can lead to problems of correlation with oxygen isotope stratigraphy. Nevertheless, such deposits provide valuable insights into local environmental change and its intensity.

Many continental archives cover only relatively short periods of the last 3×10^6 years, as is the case for many lake sediments. Occasionally, buried lacustrine sequences are discovered which relate to past interglacials; these are important because they provide information on lake development, which is unaffected by human activity. Similarly, short lacustrine sequences, peats, palaeosols, carbonate deposits and packrat middens represent limited time periods. However, the application of techniques such as pollen analysis, plant macrofossil analysis, oxygen isotope determination and absolute age determination, such as radiocarbon and uranium series, has resulted in a vast range of palaeo-environmental information that facilitates the analysis of small- and large-scale environmental change.

From this 'patchwork' of data, it is becoming possible to establish an increasingly detailed scenario of environmental change for the last 3×10^6 years. There are many other techniques of analysis that can be used to reconstruct palaeo-environments and additional archives, as will become apparent in Chapters 7, 8 and 9.

Summary Points

- There are many continental archives of environmental change.

- The longest, most complete sequences are those from middle and low-latitude lake basins and **loess** deposits.

- Many archives can be correlated with marine **oxygen isotope stages**.

- A wide range of techniques can be applied to palaeoenvironmental archives; one of the most common of these is pollen analysis.

- Multidisciplinary approaches to archive interpretation are particularly valuable.

- All techniques of palaeoenvironmental analyses have disadvantages as well as advantages.

- The age determination of archives is important to establish correlations between archives and to establish rates of environmental change.

- Data from archives may provide information on local environmental change, which in turn contributes to a regional and/or global scenario of environmental change.

General further reading

Glaciers and Glaciation. D.I. Benn and D.J.A. Evans. 1997. Arnold, London.

Global Environmental Change. A Natural and Cultural Environmental History. A. M. Mannion. 1997. Longman, Harlow, Essex, 2nd edn.

Palaeoenvironmental Events During the Last 15000 Years. Regional Syntheses of Palaeoecological Studies of Lakes and Mires in Europe. B.E. Berglund, H.J.B. Birks, M. Ralska-Jasiewiczowa and H.E. Wright (eds). 1996. John Wiley and Sons, Chichester.

Past Glacial Environments. Sediments, Forms and Techniques. J. Menzies (ed.). 1996. Butterworth Heinemann, Oxford.

Quaternary Environments. M.A.J. Williams, D.L. Dunkerley, P. De Deckker, A.P. Kershaw and T. Stokes. 1998. Arnold, London, 2nd edn.

Reconstructing Quaternary Environments. J.J. Lowe and M.J.C. Walker. 1997. Longman, Harlow, 2nd edn.

The Ice Age World. B.G. Andersen and H.W. Borns. 1994. Scandinavian University Press, Oslo, Copenhagen and Stockholm.

The record of environmental change in tree rings and historical and meteorological records

6.1 Introduction

In addition to the archives of palaeoenvironmental information referred to in Chapter 5, there are numerous other archives that provide valuable data. Such records are spatially and temporally fragmented but nevertheless provide the means for the reconstruction of local and sometimes regional environmental change. While many of the archives considered in Chapter 5 provide evidence for environmental change over relatively long time periods, i.e. 5K years or more, tree rings and documentary records provide evidence for environmental change during shorter periods, especially during the historic period.

Dendrochronology (the use of tree-ring sequences for dating) was developed in the 1960s and 1970s. It is an incremental method of age estimation, rather like the varve sequences (Section 1.2.2) in lake basins. Tree rings, however, show considerable variation between years because their width, density and composition reflect the prevailing environmental conditions. Consequently, tree rings provide not only a means of age determination, but also palaeoclimatological and palaeoenvironmental information. Whilst the longest tree-ring chronologies extend back *c*. 10K years, the tree-ring sequences used for the reconstruction of climate are generally concerned with the last millennium, as discussed below.

Historical records, in contrast, are many and varied. Accounts of sea voyages, accounts of explorers about newly discovered lands, weather diaries, estate inventories, and art (including museum pictures), etc., all provide information on past environments. Similarly, records of droughts, floods and famines provide information on catastrophic events and their periodicity, as do records of dates for the first flowering of certain plants, e.g. cherry blossom in Japan. Such records are proxy records for past climatic conditions. Such conditions can be assessed directly through recorded meteorological data, but, in view of changes in instrumentation over time, this is not always a straightforward task. Systematic records are a relatively recent occurrence, with sequences of data extending back *c*. 400 years at most, e.g. in Europe, but with much shorter sequences of data for most of the rest of the world. Such data are, however, very important in terms of identifying both natural and anthropogenic climatic change.

6.2 Tree rings

The cells that trees produce to transport water and nutrients accumulate around the edge of the trunk. Thus, older trees have a greater girth than younger trees growing under the same ecological and environmental conditions. Patterns of growth, i.e. cell accumulation, vary between species. Coniferous species and so-called ring-porous hardwoods such as oak and elm show distinct differences between cells produced during the spring when growth, promoted by increasing warmth, is greatest, and those produced during the summer and autumn, when growth slows. This seasonal variation gives rise to an annual ring, comprising large, thin-walled cells and small, relatively thick-walled cells. As Schweingruber (1988) has reported, the species most widely used in dendrochronological/dendro-ecological work, because of their distinctive tree-ring patterns, are pine (*Pinus*) and oak (*Quercus*). In the case of many hardwoods, e.g. beech, alder and lime, the pattern of tree-ring production is not as distinct (these are so-called diffuse-porous genera), so they are not so useful in palaeoenvironmental research.

The value of tree rings in palaeoenvironmental research lies not only in the capacity of tree-ring sequences to provide a dating control, but also because variations within tree-ring sequences reflect the prevailing environment. For example, periods of environmental stress such as drought, floods, etc. will affect growth rates and cell characteristics. During such periods, growth will not be as great as it is when there are no environmental stresses. Consequently, tree-ring sequences provide a record of past climatic variations on a year-to-year basis; such a fine record can rarely be obtained in other circumstances. The various procedures used in dendrochronology/dendroecology are described briefly in Box 6.1.

In relation to dendrochronology, it is possible to construct fixed and floating chronologies. The fixed chronologies refer to fixed time periods or an absolute time scale. Beginning with living trees, whose ring sequences are recorded by coring (Box 6.1), it is possible to cross-match a large number of sub-fossil wood specimens of overlapping ages, in order to establish a long chronology. To date, the greatest fixed chronology established is that of Becker and Kromer (1993), which extends back to 9971 calendar years BP and is derived from German oak remains recovered from peat and gravel deposits. In Ireland, bog oaks have facilitated the construction of a 7272 calendar-year chronology (Pilcher *et al.*, 1984).

Floating chronologies, in contrast, are, as the name implies, not linked absolutely to a particular dated horizon in sub-fossil or living wood. Nevertheless, even a tree-ring sequence from one sub-fossil trunk provides an internal chronology and it is often possible to build up fairly long chronologies with overlapping specimens. These may then be related to fixed chronologies through independent means such as radiocarbon age estimation (Box 5.2). For example, the long, fixed

Box 6.1

The principles involved in dendrochronology/dendroecology

1 To construct a tree-ring chronology which is 'fixed' in time, it is necessary to establish the ring sequence of a living tree. In Britain or Europe this would normally be oak *(Quercus)* or pine *(Pinus)*. The ring sequence can be established by extracting a core from a living tree using an increment small-diameter borer. Subsequently, samples are dried, polished and mounted so that the ring sequence can be examined by the naked eye or using a binocular microscope. Alternatively, X-ray densitometry may be used. This involves the X-raying of wood sections to produce negatives. These can then be scanned using a beam of light and a photo-cell. The amount of light passing through the negative relates to the density of the wood. The resulting measures of density reflect climatic/environmental change just as do tree-ring widths measured conventionally (see above).

2 To extend the record back in time, cross-matching of either widths or density patterns is necessary. Initially this must be between the pattern from the living tree and that from an ancient tree, e.g. a beam from a house. Then cross-matching between successively older timbers is necessary. Eventually, the cross matching may involve timbers from archaeological monuments/sites and fossil tree remains such as bog oaks and pines. The process is illustrated below:

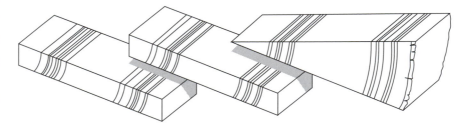

In this way, long chronologies can be constructed (see text for details). Most importantly, a master chronology for a given region can be constructed by matching the ring patterns from numerous trees for each of the 'time slices' represented. Consequently, an 'average' pattern of tree ring-width and/or density can be constructed. It is then possible to relate isolated finds of timbers to this master chronology in order to ascertain their position.

3 Various computer techniques have been developed to improve the accuracy of cross-matching/cross-dating. In addition, various statistical procedures can be used to produce ring-width indices. These allow corrections to be made to compensate for variations in growth rate that may occur between, for example, old and young trees.

For details see Baillie (1995), Cook and Kairiukistis (1990).

German oak chronology referred to above has been extended back to 11 370 calendar years BP (Becker, 1993), by relating it to a floating pine chronology (further revisions have also been undertaken; Section 6.2.1). Another example of a floating chronology is that reported by Kuniholm *et al.* (1996) for Eastern Anatolia, Turkey. Here, wood and charcoal from numerous archaeological sites have facilitated the construction of a 1503 year floating chronology that falls within the period 2220–718 BC. In addition to the significance of this chronology, which has been tied to an absolute time scale by radiocarbon age determination, for dating important cultural events, cooling events in the seventeenth and twelfth centuries BC were recognised. These can be correlated with volcanic eruptions, which caused short-term climatic cooling and which are registered in other palaeoenvironmental archives such as ice cores (Chapter 4). In particular the earlier eruption is considered by Kuniholm *et al.* to be that of the volcano Santorini/Thera in the Aegean Sea. This occurred in 1628 BC, and evidence for it is found as far afield as North America, reflecting its global impact. This study highlights the significance of dendroclimatology/dendroecology for archaeological studies.

6.2.1 Long-term tree-ring sequences

As stated above, one of the longest fixed dendrochronologies so far obtained is that described by Becker and Kromer (1993), based on bog oaks from Germany. The correlation between this and an earlier but floating chronology based on pine remains (Becker, 1993), has been revised by Bjorck *et al.* (1996), see above, and now extends to 11 450 years BP. This dendrochronological record also provides a dendroclimatological record; both have been related to lake sediments from Sweden and ice-core records for the period 11.5K years to 8.5K years BP. A summary of the tree-ring and ice-core oxygen isotope data (Section 4.3.1) is given in Figure 6.1. The results of the correlation indicate that the Younger Dryas climatic regression occurred between 12.6K and 11.45K calendar years BP. In addition, Bjorck *et al.* recognise a cooling episode of *c.* 150 years duration in the early postglacial period. This cooling episode began *c.* 11.2K years ago and is referred to as the Preboreal oscillation. During this period, tree-ring widths increased (Figure 6.1), because of enhanced humidity; this contrasts with increased aridity, beginning at *c.* 11K years BP, when tree-ring widths declined. This study illustrates the fine resolution that is now possible for the reconstruction of past environmental change and the merits of a multidisciplinary approach.

Figure 6.2 gives a further example of a relatively long tree-ring chronology representing the last *c.* 3500 years. This record was derived from living and sub-fossil alerce (*Fitzroya cupressoides*, which is also known as the Patagonian cypress) remains from south-central Chile. In general, the data reflect a relatively stable climate for most of this period, with three intervals of

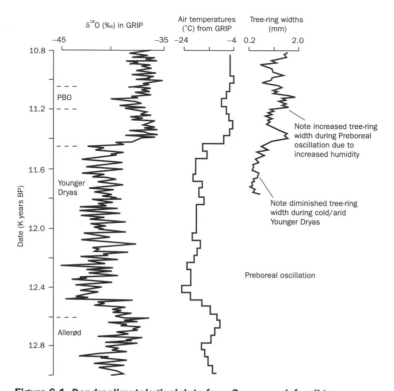

Figure 6.1 *Dendroclimatological data from German sub-fossil tree remains in relation to oxygen isotope data from the GRIP ice core*

Source: Based on Björck *et al.* (1996).

prolonged above or below-average temperatures. These are indicated on Figure 6.2. The latter period of below-average temperatures is the Little Ice Age, a period of widespread cooling, which is recorded in a variety of palaeoenvironmental archives. This period is discussed in detail in Sections 6.2.2 and 6.3.

In the south-western region of the USA, the bristlecone pine (*Pinus longaeva*) is an important component of the flora, and occupies arid and often rocky habitats up to 4000 m. These trees are particularly significant in dendrochronological studies because they are exceptionally long lived when compared with other tree species, and because they are particularly sensitive to climatic change.

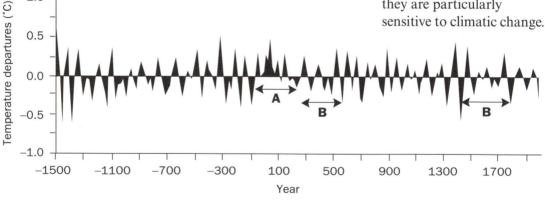

A: longest interval with above average temperatures

B: longest intervals with below average temperatures

Figure 6.2 *Reconstruction of the departure of summer temperatures from average values (December to March) from a south-central Chile tree-ring record for the last 3.5K years*

Source: Based on Lara and Villalba (1993).

Some specimens are more than 4000 years old and so provide a long living tree-ring record against which sub-fossil material can be compared. The cross-matching of living, dead and sub-fossil material has facilitated the construction of a fixed 8681 year chronology (Ferguson and Graybill, 1983).

6.2.2 Short-term tree-ring sequences

The majority of tree-ring records relate to the last 1000 years or so. In many cases it has proved possible to relate the dendroclimatological record to written records of weather and climatic events. There are many examples of such tree-ring sequences from many parts of the world (see reviews in Lowe and Walker, 1997; Mannion, 1997b).

Many studies extend back to *c.* 1500/1600 and so provide a record of relatively recent climatic trends, including periods of drought and temperature declines caused by volcanic activity. One such example is a study by Case and MacDonald (1995) to establish a fixed chronology based on living and sub-fossil limber pine (*Pinus flexilis* James) from the western edge of the Great Plains in south-western Alberta. Starting with living trees as old as 526 years, and an established relationship between ring growth and annual precipitation, Case and MacDonald produced a 487-year record of annual precipitation for the region. This is the longest record so far produced and it shows that drought occurred periodically. A comparison between the last 100 years of the dendrochronological record with instrumental records shows a similar incidence of drought. This indicates that the dendroclimatological record is valuable in its own right. Moreover, on the basis of the pre-instrumental record for the last 500 years from Alberta, there was no earlier episode of drought in the later Holocene that was more severe than that of the 1918–1922 period, which even led to the abandonment of farms.

A 400-year record of rainfall based on tree-ring density and width has been reconstructed for north-central China (Hughes *et al.*, 1994) using a species of pine (*Pinus armandii*). A substantial drought occurred in the 1920s, as it did in Alberta, Canada (see above), and the results can be corroborated by documentary evidence. An earlier severe drought occurred in the early 1680s. In addition, Jones *et al.* (1995) have collated tree-ring records from some 97 sites in North America and Europe and identified a series of cool summers in both continental records. These occurred in 1601, 1641, 1669, 1699 and 1912, of which four correspond with documented, major volcanic eruptions. This study indicates that major volcanic eruptions can have a widespread, possibly global, impact on annual temperatures.

Another example of a short-term tree-ring record is that of Jacoby *et al.* (1996) for Central Mongolia (altitude 2450 m) based on Siberian pine (*Pinus sibirica* Du Tour). This record extends back 450 years to *c.* 1550, as illustrated in Figure 6.3. Of particular note are the cool periods centring on 1600, the early 1700s, cooling

A.

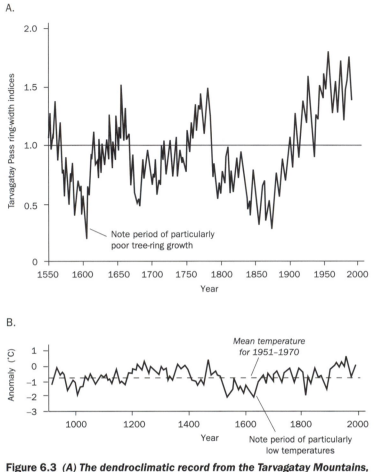

B.

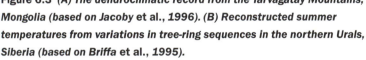

Figure 6.3 *(A) The dendroclimatic record from the Tarvagatay Mountains, Mongolia (based on Jacoby* et al., *1996). (B) Reconstructed summer temperatures from variations in tree-ring sequences in the northern Urals, Siberia (based on Briffa* et al., *1995).*

during the 1800s, and then a substantial warming trend which began in the late 1800s. Many of these variations have been recorded in other palaeoenvironmental archives such as the Dunde ice cap of China and tree-ring sequences from the Arctic, though the correlation is best for the post-1700 period. The Mongolian tree-ring sequence also contains evidence for a sustained period of warming in the last 100 years. This has no earlier counterparts and may reflect anthropogenic warming due to emissions of heat-trapping gases and the enhancement of the greenhouse effect. A further record from Siberia (Briffa *et al.*, 1995) shows similar trends to that from Mongolia (shown in Figure 6.3). In particular, the temperature depression in the early 1600s can be identified, as can a run of below-average years in the 1800s and the sustained warming trend of the last 100 years. High temperatures since the 1970s have also been recorded in Patagonian tree-ring sequences, though there are earlier, pre-industrial periods of equally high-temperature anomalies (Villalba *et al.*, 1997).

Much research has also focused on the Little Ice Age. This is a difficult period to define, because the evidence for a widespread and possibly global period of cooling that led to glacier and ice-cap growth is time transgressive (see Grove, 1988). In high and middle latitudes the evidence for ice advance is considered to have occurred in the 1400s, 1500s and 1600s. According to the synthesis of Bradley and Jones (1993), the coldest conditions occurred between 1570 and 1730, and also in the early 1900s before the onset of a warming trend in the 1920s. Some

Table 6.1   *Climatic reconstructions for Patagonia, based on tree-ring data*

Cold/warm intervals	Drought/wet intervals	Glacial advances
	D 1770–1850	
C 1270–1660		
Little Ice Age	D 1570–1650	1520–1670
	W 1450–1550	
	D 1280–1450	1270–1380
W 1080–1250 Equivalent to the Medieval Warm Epoch?	W 1220–1280	
C 900–1070		

Source: Based on Villalba (1994).

agreement with this comes from Villalba's (1994) work on Patagonian tree-ring sequences, the information from which is summarised in Table 6.1. This shows that during the Little Ice Age there was no uniformity of climate; wet and drought periods alternated and there were two major episodes of glacial advance. There is also a considerable volume of information on the Little Ice Age from historical and meteorological records, as is discussed below.

6.3 Historical records

There is a great variety of historical records that have been used as proxy records of past climatic characteristics. Such records provide evidence of local and, possibly, regional variations in climate, though these may reflect natural climatic variability rather than sustained climatic change. Examples of historical records used in this way include observations on extreme weather events such as storms, droughts and floods, records of the frequency of spring frosts and records of the phenological rhythms of certain plants, such as the dates of the first flowering of the cherry blossom in Japan. Moreover, artists and authors have frequently left behind a painted or written record in their works of past weather conditions in various parts of the world. Since the 1800s, photographs have come into existence and these may also be used as evidence of past climatic characteristics.

6.3.1 Drought/flood records

China has a long history of meteorological observations and these have been investigated by chroniclers of climatic history in order to reconstruct past climates and their impact. A recent example of such work is that of Currie (1995), who has analysed 81 long and 202 short dryness/wetness indices which have been compiled from more than 2000 sources, including local annals, by the Central Meteorological Institute of China. The dry and very dry indices compiled are given in Figure 6.4. Several periods of drought can be recognised; for example,

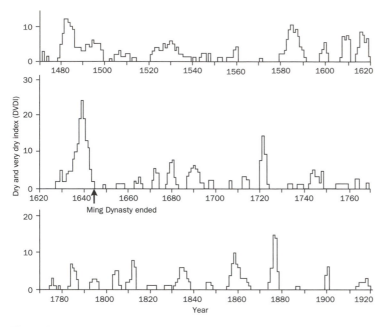

Figure 6.4 *Dry and very dry indices (DVDI) of climate in China, as derived from historical records*

Source: Based on Currie (1995).

between 1480 and 1500 there was a prolonged period of above-average dryness with another not so long but particularly acute drought between 1633 to 1644. It is interesting to note that the Ming Dynasty fell in 1644, just as this period of drought ended. Perhaps the emperor and the land-owning élite could not withstand the onslaught by the peasants as their crops failed and famine ensued. It is equally interesting to note that tree-ring records from a number of localities also reflect drought around this time, as discussed in Section 6.2.2 above.

Jiang *et al.* (1997) have also used indices of wetness derived from historical records dating back to AD 960 for parts of East China. Using a statistical method known as the 'Mexican hat' wavelet technique, they have reconstructed temporal and spatial rainfall variability in six regions. In addition, the data indicate a persistent dry stage between 1120 and 1220, a wet stage between 1280 and 1390 in the coastal regions, and a shift from a weak wet phase to a weak dry phase in East China generally, since 1890.

Documentary evidence of an ecclesiastical nature has been exploited by Barriendos (1997) to reconstruct climatic patterns in Spain during the period 1675–1715, part of the Little Ice Age. Using ecclesiastical chapter records, which contain a wealth of information on local events, crop harvests, etc., Barriendos focused on rogation ceremonies (for the three days before Ascension Day). These are religious ceremonies at which supplications and requests are made. In particular, Barriendos examined the record of *pro pluvia* and *pro serenitate* vocations, i.e. requests for rain and for 'fair weather'. The data show that during the 1675–1715 period there was an overall increase in precipitation on the Iberian Peninsula, with marked regional variation. This is attributed to an increase in the frequency of meridional circulation at the expense of zonal circulation.

Weather conditions, including the occurrence of drought and summer rainfall, for Crete during the years 1547/48 to 1644/45 have been reconstructed by Grove

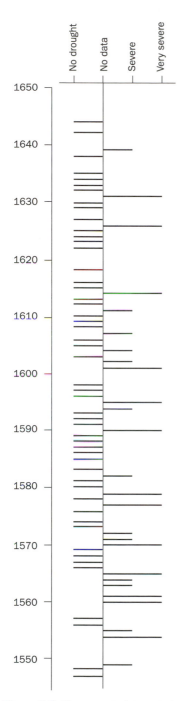

Figure 6.5 *The record of drought in Crete 1547/48 to 1644/45, as reconstructed from documentary evidence*

Source: Based on Grove and Conterio (1995).

and Conterio (1995). This was achieved on the basis of a variety of documentary evidence for the period in question and via a comparison with twentieth-century climatic characteristics. The documentary sources are many and varied, and include correspondence between officials in Crete and the Doges of Venice, who ruled the island between 1204 and the late 1600s. Such documents are today lodged in the Venetian State Archives. They provide proxy climatic data in the form of reports on crops, dry spells, damage caused by high winds or heavy rains, etc. Figure 6.5 gives the reconstructed record of drought; it shows that a high incidence of drought characterised the periods 1561 to 1566 and 1600 to 1607, when winds came from the south as air masses extended beyond the Sahara. The latter period corresponds with the Little Ice Age, and was one in which other weather anomalies, relative to the climate of the twentieth century, occurred.

6.3.2 Other documented proxy data

Many studies directed at the reconstruction of climate during historical times have drawn on a variety of sources of information. As in the case of multidisciplinary palaeoenvironmental work, the use of a range of indices is preferable to the use of a single index, simply because the resulting explanations must fit a range of data from different sources. Such a multi-proxy approach has been employed by Pfister (1992), who has used some 33 000 records involving a combination of weather diaries and weather reports by clergy, dates recorded for the establishment and melting of snow cover, and the dates of flowering, etc. for various crops and plants. These proxy data have facilitated the reconstruction of monthly temperatures and precipitation in Central Europe for the period 1525 to 1979. The conclusions drawn by Pfister include the fact that prior to 1900, winter and spring months were colder and drier than those for 1900 to 1970. In addition he showed that climatic variability was greater before 1900 than after, with pronounced variability around 1600. These data corroborate those cited in Sections 6.2.2 and 6.3.1 for extreme conditions in the early 1600s. Pfister's data also show that a warming trend has been established in recent years. This again mirrors tree-ring data from numerous localities such as Siberia and Mongolia (see Section 6.2.2).

Using a similar approach, Pfister *et al.* (1996) have examined the severity of winter temperatures in Central Europe and Northern Italy. They collated some 2133 records, many originating from the *Monumenta Germaniae Historica*, including records on snow-cover, ice extent and crop/plant phenology. Their results show that there was a period of cold winters between 1303 and 1328, with the winters characterised by average temperatures between 1329 and 1354, and then a period of winters with variable temperatures between 1354 and 1375; until 1400, winters generally experienced below-average temperatures. Another, similarly cool period is reflected in proxy climatic data from Western Europe between 1675 and 1704 (Wanner *et al.*, 1995). Cool and dry conditions prevailed, with an increased extent of sea-ice and low ocean temperatures. Why such 'runs' of cool/cold conditions should have occurred is, however, a matter for debate.

Newspaper accounts can also yield valuable information on past weather conditions, especially extreme events. This has recently been demonstrated by Mason *et al.* (1996), who have analysed reports in two local Alaskan newspapers dating back to 1899. The data show that, in the Bering Sea, storm surges were rare from 1916 to 1928, and from 1947 to 1959; the most frequent and intense storms occurred from 1900 to 1913, from 1936 to 1946, from 1974 to 1976, and again in 1992.

6.3.3 Evidence for climatic change from art and photography

Lamb's (1995) review of documentary and other proxy sources for the reconstruction of climate shows that there is a considerable volume of information available from works of art. Some of these even provide evidence for climatic change in the prehistoric period. Moreover, the advent of photography in the early 1800s provided another means of recording information, which has since become a palaeoenvironmental archive.

Some of the world's earliest cave paintings, such as those from the Chauvet Cave in the Ardeche of France, which are *c.* 32K years old, and the famous Lascaux paintings from the Dordogne dated at 17K years BP, attest to the presence of a cold and tundra-like environment quite different to the present-day environment in these regions. Similarly, as Lamb (1995) highlights, there are rock paintings from the Sahara desert, some 8K years old; these attest to the presence of surface water in some areas, which supported species such as buffalo, elephants and crocodiles. Certainly, these species are no longer present in this arid region today. Records like these can often be substantiated by palaeoenvironmental and archaeological data.

More recent works of art, however, provide pictorial evidence of climatic characteristics for specific locations, and they are usually precisely dated. Lamb (1995) draws attention to the Little Ice Age in Europe, notably the 1500s and 1600s, and the art that depicted frozen rivers, early winter scenes and severe winter snows. He points out that the River Thames in London, today a city warmer than the surrounding countryside by several degrees Celsius because of the urban heat-island effect, froze over on at least 11 occasions in the 1600s. During this period there are records of extensive snow cover, frozen canals, crop failures and cool summers much of which is reflectied in artists' efforts to capture their environment in paint. Such artists include Peter Brueghel the Elder and Abraham Hondius.

An example of the use of photography as a tool for the assessment of environmental change is provided by Gellatly *et al.* (1994), who have been able to pinpoint specific dates when the Taillon Glacier, in the French Pyrenees, advanced or retreated. Their research involved a multi-proxy approach, using early maps and paintings as well as photographs in conjunction with actual measurements of the ice margin. They were able to determine that the glacier retreated overall, but also re-advanced briefly in the periods 1886–1890, 1906–1911, 1926–1928, 1945 and 1964. Is this overall retreat yet another indication of recent global warming, which is also indicated by the tree-ring data from Mongolia and Siberia, as referred to in Section 6.2.2?

6.4 Meteorological records

Although meteorological records are the most reliable records of past climates and climatic variability, they still have drawbacks. First, systematically collected records do not extend far back in time; for example amongst the earliest systematic records are those from England, the collection of which began in the late 1600s. Moreover, the extent of such records is variable spatially, due to the fact that systematic collection of meteorological data in many countries only began in the early 1900s. An additional problem occurs, especially with early data, in relation to compatibility between data of a similar age but collected in different places, and between early data and modern data that were collected using different instrumentation. In most cases the establishment of long sequences of data requires the use of correction factors to ensure compatibility between old and recent data. Meteorological records are particularly useful for establishing the extent of natural variability of climate. Many of the examples given in the sections above, for example, refer to wet years, drought years, periods of aridity, etc. Such periods may, however, reflect natural variability rather than the establishment of trends that become sustained over long-term periods and can thus be considered as shifts in the climate system. What constitutes change and what constitutes natural variability about a near-constant mean is central to the current debate on global warming, an issue that is discussed in Section 6.5. Records of past climate are important for distinguishing between variability and change; they also provide 'norms' or averages for precipitation, daily/monthly/annual temperature regimes, atmospheric pressure, storm incidence, etc. For example, in Figure 6.3(B) the temperatures reconstructed from Siberian tree-ring data are expressed as anomalies about a mean; this was established on the basis of meteorological data recorded between 1951 and 1970.

The establishment of what constitutes variability and what constitutes change is, however, not the only problem associated with meteorological (and other) records. Whilst meteorological records, and indeed all types of palaeoenvironmental records considered in previous chapters, provide evidence for variability and/or change, it is often impossible to determine if this is due to natural or anthropogenic causes. Since the advent of agriculture c. 10K years BP (see Mannion, 1995 for a review), environmental change has accelerated, as it has throughout history and even more so with the Industrial Revolution. Whilst such anthropogenic environmental change is of great significance (Mannion, 1997b), it adds to the difficulties of discerning natural change in environmental systems that are also dynamic. The coincidence of the advent of meteorological records and intensifying human activity means that the establishment of base-line data is problematic.

The impact of volcanic eruptions is sometimes evident in instrumental records. For example, one of the most intense eruptions in recent years has been that of Mount Pinatubo in the Philippines, in 1991. According to Parker *et al.* (1996), this had a marked impact on global temperatures for at least three years after the eruption. They report that global mean air temperatures were reduced by as much as 0.5°C at the surface and 0.6°C in the **troposphere** (the atmosphere up to c. 12 km above the Earth's surface) for several months in mid-1992. The impact was, however, relatively short lived, having diminished significantly by 1994.

One long sequence of meteorological data, from Cadiz and San Fernando in Spain, has been examined by Wheeler (1995) who has reported a 2°C cooling for the years 1805–1816. This may be due to dust produced by volcanic eruptions. Since 1816 these instrumental records indicate the onset of a warming trend. This is also evident, but with a later start in the 1880s, in tree-ring records (Section 6.2.2). However a warming trend is not in evidence everywhere. For example, Przybylak (1997) has analysed temperature data for the Arctic collected between 1951 and 1990 and found no evidence for a warming trend. In contrast, Singh (1997) has detected a temperature increase of c. 1°C and fluctuating rainfall patterns in instrumental records from Trinidad in the Caribbean over the last 50 years. Similarly, there is evidence for two warming phases in temperature records collected near Harare and Bulwayo in Zimbabwe for the period 1897 to 1993 (Unganai, 1997). The first of these occurred between the mid-1930s to 1940, and the second began in 1980 and is continuing.

On a global basis, there is evidence to suggest an overall warming trend of *c.* 0.5°C since 1900. This conclusion is reached from the analysis of temperatures collected from 5400 recording stations distributed around the world, and has led Easterling *et al.* (1997) to conclude that this increase is at least partly due to a decrease in the **diurnal temperature range** (DTR). This is the difference between daily maximum and daily minimum temperatures. Mean annual temperature departures from the 1961–1990 mean are given in Figure 6.6. This shows clear evidence for a global-warming trend since the late 1970s in both hemispheres, though the trend is most pronounced for the Northern Hemisphere. It is interesting to note that the anomalies from means are similar in overall pattern to those given in Figure 6.3 for tree-ring sequences. Equally interesting is the parallel between temperature increases indicated by the Patagonian Southern Hemisphere tree rings (Villalba *et al.*, 1997; Section 6.2.2) and actual temperature increases detected in instrumental records (Figure 6.6). The recognition by Villalba *et al.* that the warming trend in the Southern Hemisphere has earlier precedents suggests that actual temperature increases in the Southern Hemisphere may still be within the realm of natural variability, in contrast to those of the Northern Hemisphere which appear to have no earlier precedents.

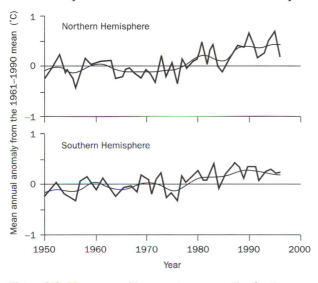

Figure 6.6 *Mean annual temperature anomalies for the Northern and Southern hemispheres, 1950–present*

Source: From Easterling *et al.* (1997).

6.5 Conclusions

The sections above attest to the varied nature of sources of evidence for past climate. In terms of proxy data sources, tree-ring sequences provide the longest data series. The longest chronologies extend back nearly 10K calendar years, with the possibility of extension to almost 11.45K calendar years. Such sequences have particular relevance for age determination and the calibration of radiocarbon dates, but are also valuable for reconstructing climatic change in detail. Moreover, dendrochronology/dendroclimatology has considerable application in archaeology, notably for establishing links between climatic change and cultural developments. There are also numerous relatively short-term tree-ring sequences (for approximately the last 1000 years), which are important for establishing climatic patterns and variability in historical times, including periods not covered by instrumental records. Several such records show a marked warming trend in the twentieth century and thus contribute to the debate on the recent enhancement of the greenhouse effect due to the generation of heat-trapping gases by human activity.

Historical records are equally important for establishing climatic patterns, etc. in the recent past. Such records occur in many different forms, most of which are proxy rather than direct records. Documentary evidence may include data on crop harvests, adverse weather records, e.g. storms, floods and droughts, dates of first flowering for plant species, e.g. vines, and religious ceremonies that were directly related to weather conditions. Letters and inventories between the conquered and the conquering, as in the case of Crete between the twelfth and seventeenth centuries, provide varied information on climatic patterns. In particular, climatic deterioration during the Little Ice Age is often evident from such sources. Art and photography also have their place in studies of past climates; they provide a visual record of flora, fauna and landforms.

The most direct records of past climate are, of course, instrumental records. These extend back only 400 years at most, and there are problems of compatibility between older and recent records. Nevertheless, in many such records there is a twentieth-century warming trend. There is much debate as to whether this is due to natural variability or whether it represents a real change to a warmer world. This highlights the difficulties involved in isolating natural environmental change from anthropogenic environmental change.

Summary Points

- Tree-ring sequences extend as far back as 11.5K calendar years.

- As well as facilitating age determination, tree-ring sequences provide information on climatic variation, including the incidence of drought and below-average temperatures.

- Many tree-ring sequences provide evidence for a warming trend that began in the early 1900s.

- There are many types of other proxy records of climatic characteristics: notably documentary records, art and photography.

- Much effort has focused on proxy records of climate during the Little Ice Age.

- There is widespread evidence for climatic deterioration during the Little Ice Age, which may have been global in extent.

- Instrumental records are the most reliable means of reconstructing climatic patterns and variations.

- There is much debate as to what constitutes climatic change rather than climatic variability, and also what constitutes natural rather than anthropogenic climatic change.

General further reading

A Slice Through Time. Dendrochronology and Precision Dating. M.G.L. Baillie. 1995. B.T. Batsford, London.

Climate Since AD 1500. R.S. Bradley and P.D. Jones (eds). 1992. Routledge, London (reprinted with revisions in 1995).

Climate, History and the Modern World. H.H. Lamb. 1995. Routledge, London, 2nd edn.

Climates of the British Isles: Present, Past and Future. M. Hulme and E. Barrow (eds). 1997. Routledge, London.

The Little Ice Age. J.M. Grove. 1988. Methuen, London.

7 Environmental change in high latitudes (latitudes 60–90°N and 60–90°S)

7.1 Introduction

Whilst the last 3×10^6 years have been characterised by a dynamic environment globally, high latitudes have been particularly severely affected because they have experienced the direct impact of glacial advance. As is demonstrated by ocean-sediment records (Chapter 3), there were many cold stages during the last 3×10^6 years, possibly as many as fifty. Consequently high latitude regions, i.e. those regions north of 60°N and south of 60°S were actively glaciated for numerous periods, each lasting c. 100K years. Today, the high Arctic and Antarctic zones are experiencing glaciation and thus provide modern analogues against which to assess past processes and conditions.

Most of the evidence for high-latitude environmental change derives from the Northern Hemisphere. This is because of the greater extent of land area compared with the same latitudes in the Southern Hemisphere. Moreover, research on glacial deposits and processes traditionally has been most intense in Northern Hemisphere countries (Section 1.2). In recent years, however, ocean-sediment cores from the Southern Ocean have contributed to the current understanding of environmental change in the region. In combination with ice-core data from the Antarctic (Chapter 4), and increasing evidence from a variety of sources in southern Argentina and Chile, it is becoming possible to reconstruct some of the detail of environmental change.

Indeed, in both the Northern and Southern hemispheres the most complete record of environmental change is present in ocean sediments, with data from ice cores augmenting this record and, in some cases, notably the Greenland ice cores, creating considerable controversy. This is itself beneficial because it promotes research and calls results into question. It should also be noted that research in high latitudes is particularly important because of the predicted high susceptibility of such regions to global warming. Examining the impact of past periods of warming (and cooling) will contribute to the refinement of predictive models, and thus facilitate future planning.

7.2 Greenland and the polar region

The glaciated region above 60°N comprises a permanent ice pack over much, but not all, of the Arctic Ocean, a substantial ice sheet over Greenland, and permanent small ice caps on several islands such as Baffin Island, Svalbard, Iceland and Novaya Zemlya (Russia). In this region, the most complete record of environmental change in the last 3×10^6 years derives from ocean sediments (Chapter 3). The limited extent of landmasses, when compared with the middle and low latitudes, and the repeated advance and decay of ice sheets during this time, means that the terrestrial deposits here are limited in extent. They also tend to represent relatively recent events, as the most recent ice advance destroyed evidence of earlier **glacials** and **interglacials**. However, whilst the advancing ice can be destructive, the zone of accumulation can eventually become an archive of environmental change, as discussed in Chapter 4. The deepest ice cores (representing the longest time period) so far extracted from Greenland, are the GRIP and GISP2 cores, which are referred to in Chapter 4. These represent the last *c.* 180K years.

The most complete record of environmental change in the Arctic derives from ocean-sediment cores. For example, Spielhagen *et al.* (1997) have examined the record of glacial–interglacial changes in a sediment core from the Lomonosov Ridge in the Central Arctic Ocean (Figure 7.1). They have focused on the composition of ice-rafted sediments to determine the source areas in nearby continental landmasses. They show that between 2.8×10^6 years BP and 0.7×10^6 years BP the material deposited comprised lithic carbonates with a source area in North America. After 0.7×10^6 years BP, the ice-rafted debris is dominated by smectite, which is characteristic of northern Siberia. In addition, four glacial stages occurred during this period, reflecting intensified continental glaciation in Siberia relatively late in the Quaternary period when compared with the pre-Quaternary onset of glaciation in North America *c.* 2×10^6 years earlier.

The last 1×10^6 years of environmental change have also been reconstructed on the basis of sediment cores from Northwind Ridge, Amerasia Basin (Phillips and Grantz, 1997). A summary of the conclusions from this study is given in Figure 7.2. The stratigraphy itself reflects the oscillation of glacial and interglacial conditions: sediments deposited during glacial stages comprise grey **pelagic** (deep-ocean) **muds** in which fossils and microfossils are largely absent; in contrast, sediments deposited during interglacial stages comprise orange–red muds, which reflect oxidising conditions, and which contain fossils of various kinds. The grey sediments were deposited as the sea-ice sheets covered the Amerasian Basin, and resulted in reducing conditions that were not conducive to photosynthesis. The stratigraphic record indicates that the formation and disintegration of such ice sheets occurred relatively rapidly. On disintegration, surges of icebergs were able to enter the Amerasian Basin, where they deposited erratics and sediments. Moreover, oxidising conditions returned as an increased amount of oxygen was

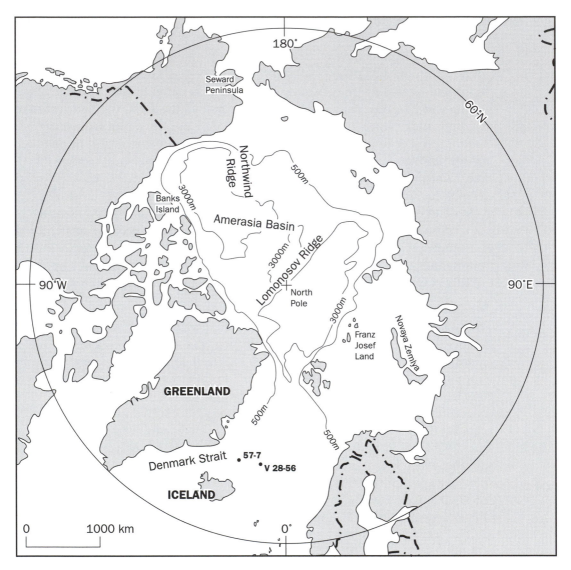

Figure 7.1 *The Arctic region today and the location of sites referred to in the text*

Note: 57–7 and V 28–56 are marine sediment cores

able to penetrate the water body; the increased availability of light also stimulated photosynthesis. The distribution of erratics in interglacial sediments reflects the fact that during many of the interglacial stages northern North America continued to be glaciated and contributed icebergs, and hence deposits of erratics, to the Amerasia Basin. This glaciation was, however, not as severe or extensive as it became during the glacial stages. The periodicity of environmental change in these cores is *c.* 100K years, in agreement with other ocean-sediment records (Chapter 3) and represents an affirmation of the impact of Milankovitch cycles (Section 2.4; Box 2.2).

		Environment	Marine oxygen isotope stages		Age estimate (K years BP)
Quaternary	Upper Pleistocene	Holocene — Interglacial	1	A	10
		Glacial	2, 3, 4	B	
		Interglacial	5	A	
		Glacial	6	B	
		Interglacial	7	A	
		Glacial	8	B	
		Interglacial	9	A	
		Glacial	10	B	
		Interglacial	11	A	
		Glacial	12	B	
		Interglacial	13	A	
		Interglacial	15	B	
		Glacial	16	A	
		Interglacial	17	B	
		Glacial	18	A	
		Interglacial	19	B	
		Glacial	20	A	
		Interglacial	21	B	783
	Lower Pleistocene		22/23	A	990
		Glacial / Interglacial			
		Jaramillo Polarity subzone: condensed glacial/interglacial cycles			1070
					1190

A: Sediments of interglacial stages, i.e. orange-red muds formed under oxidising conditions and containing fossils, which provide evidence of enhanced productivity.

B: Sediments of glacial stages, i.e. grey muds formed under reducing conditions; these have a low or no fossil content, reflecting low productivity.

Note: Odd numbers denote warm/interglacial stages; even numbers denote cold/glacial stages

Figure 7.2 *A summary of the stratigraphy and the palaeoenvironmental inferences from sediment cores extracted from Northward Ridge, Amerasia Basin*

Source: Based on Phillips and Grantz (1997).

The oscillation of glacial and interglacial stages is well represented in ocean sediments from the Arctic, as the research referred to above attests. However, there is also evidence to show that both the glacial and interglacial stages were similar in terms of overall trends but different in terms of detail. The differences that occurred during glacial stages are exemplified by those recorded in the core from the Lomonosov Ridge (see above), whilst the differences occurred during the interglacial stages are illustrated by those recorded in a core (57-7) from the Iceland Sea at *c.* 68°N (Eide *et al.*, 1996). The location of the coring site is shown in Figure 7.1, and the core itself contains sedimentary material relating to **oxygen isotope stages** 1–11. On the basis of foraminiferal and coccolith assemblages (Sections 3.2.3; 3.2.5), Eide *et al.* determined that each interglacial stage was characterised by water masses of different types. A summary of their results is given in Table 7.1, which shows that the interglacial represented by oxygen isotope stage 5e was exceptionally warm in relation to the preceding

Table 7.1 *A comparison of conditions in the Iceland Sea (based on core 57-5) during the last four interglacial stages*

Oxygen isotope stage	Water-mass type	Water temperature	Productivity
5e	Warm Atlantic water from the Norwegian Current and possibly the Irminger Current	4°C warmer than the Holocene	High but variable, reflecting climatic fluctuations *
7	Low inflow of Atlantic water as compared with 5e and the Holocene	Sea-ice may have been present for part of each year	Lower than in 5e and the Holocene
	Arctic water significant	Temperatures lower than 5e	
9	As stage 7	As stage 7	As stage 7
11	Arctic rather than Atlantic water predominated. The latter entered the Iceland Sea via the Norwegian Current, but this was considerably weaker than in stage 5e	Possibly warmer than the Holocene and stages 7 and 9, and similar to stage 5e	High productivity, possibly higher than stage 5e

* This finding has parallels with oscillations recorded in the GRIP ice core (Section 4.3.1).

Source: Based on Eide *et al.* (1996).

three interglacial stages and the subsequent Holocene.

The oscillations characteristic of the climatically complex period at the end of the last ice age are registered in many archives of environmental change in the Arctic region. For example, the shifts from cold to warm and back to cold conditions, prior to the opening of the Holocene, are recorded in ocean sediments, Arctic ice cores and lacustrine sediments. In particular, Figure 4.10 gives the temperature reconstruction for the period 18K calendar years BP to 10K calendar years BP, from the GISP2 ice core. It shows a substantial shift in temperature as the last glacial maximum ended and abrupt warming began at *c*. 15K calendar years BP. This was a relatively short-lived warming period, as temperatures then began to decline at *c*. 14K years BP and reached lows between 11.2K years and 13K calendar years BP. This is known as the **Younger Dryas**, which is a cooling event recognised in many parts of the world. A subsequent rapid warming trend began at *c*. 11K calendar years BP (data are from Kapsner *et al.*, 1995). This was the beginning of the Holocene. A similar sequence of events is registered in many other Greenland ice cores (Johnsen *et al.*, 1992). Moreover, ice-core data indicate that the magnitude of temperature change between glacial conditions and the Holocene may have been as much as 15°C, with the coldest periods of the glacial stages being *c*. 21°C colder than at present (Cuffey *et al.*, 1995).

During the Holocene itself, tundra vegetation colonised the land areas as temperatures ameliorated and ice retreated north. In general, climate has remained relatively stable during the Holocene, except for the Little Ice Age *c*. 1600s to the early 1800s. The climatic deterioration that occurred during this period caused Greenland to be cut off by the Arctic sea-ice that occupied the Denmark Strait between Iceland and Greenland. In Greenland itself, and in Iceland, glaciers re-advanced and the advent of cold Arctic water led to the demise of the cod-fishing industry (Grove, 1988; Lamb, 1995).

In a recent report, Overpeck *et al.* (1997) have reviewed a range of evidence for environmental change in the Arctic region during the last 400 years. They have compiled data from tree rings, lake sediments, marine sediments, records of ice melt, and instrumental records from a wide range of locations throughout the region, in order to reconstruct climatic variability. The conclusions reached from this synthesis include the fact that warming began in about 1840 and brought the Little Ice Age to an end (see above). The warming trend continued into the mid-1900s and prompted the melting of glaciers, permafrost and sea-ice, as well as changes in lacustrine and terrestrial ecosystems. In addition, Overpeck *et al.* suggest that the warming was between 1°C and 3°C with an average for the region of 1.5°C. However, in the pre-1920 period they believe that the warming trend was a response to natural factors, notably increased solar irradiance and decreased volcanic activity. After 1920, warming due to greenhouse-gas emissions occurred to enhance 'natural' warming and/or to mask natural variability. This is characteristic of the region as a whole, but there are also marked variations within it. Overall, and taking into account a reduction in the natural warming trend since *c*. 1840, the region has warmed *c*. 0.6°C since 1900, as indicated by instrumental records. The implications of these results are that the modelling and prediction of future climatic change are likely to be difficult and may be inaccurate.

7.3 Scandinavia

Figure 5.3(B) illustrates how much of Scandinavia lies above 60°N, and shows that almost all of the region was covered by ice during the last three major glaciations. Consequently, reconstructing Scandinavia's environmental history for the last 3×10^6 years from the terrestrial record alone is problematic; the fragmented record created by the retreat and advance of ice makes interpretation difficult, as it does elsewhere in high latitudes. In Scandinavia there is unequivocal evidence for at least three major ice advances, i.e. those referred to as the Elsterian, Saalian and Weichselian (in accordance with the terminology used for Northern Europe, which is shown in Fig. 2.4). There is tentative evidence for an earlier ice sheet, which is equivalent to a pre-Cromerian glaciation (probably oxygen isotope stage 14: Fig. 2.4). However, evidence from ocean-sediment cores from the North Atlantic Ocean indicate that there were many additional glacial and interglacial stages. Some of this evidence has been discussed in Chapter 3.

In Scandinavia itself, evidence for stages other than the last ice age (the Weichselian) is found in areas peripheral to the maximum extent of the ice sheet (Figure 5.3B). This has been reviewed by Donner (1995), who points out that the majority of sediments of **interglacial** ages lie well to the south of 60°N, notably in Jutland. The earliest is that of Harreskov which is considered to be equivalent to interglacial II of the Cromerian Complex, i.e. approximately 687K years to 627K years BP (Figure 2.4). However, there are numerous sites of Eemian (last) interglacial age in northern Sweden and Finland at roughly latitudes 65–67°N. These are freshwater sediments containing pollen assemblages. At Leveaniemi, for example, the pollen assemblages indicate the development of a boreal-forest type vegetation dominated by birch, pine, spruce, alder and hazel. The remains of certain beetles in the sediments indicate that the Eemian was a warmer interglacial than today, as is suggested by many other types of evidence throughout Europe and elsewhere.

Following extensive glaciation during the last (Weichselian) ice advance (Figure 5.3B), Scandinavia experienced considerable oscillations of climate as the ice sheet waned. This is reflected by changes in sedimentary characteristics and foraminifera (Figure 3.3) in the northern Norwegian Sea, as summarised in Table 7.2. The **Younger Dryas** climatic deterioration, for example, was marked by both

Table 7.2 *Late glacial and Holocene environmental change, as indicated by marine sediments from the northern Norwegian Sea*

Age (K years BP)	Dominant foraminifera	Palaeoenvironment
7.8–present	Cassidulina teretis Cibicides wuellerstorfi	Surface temperature declined to a low at c. 6.5K years BP, but then increased to reach a maximum at c. 2K years BP due to an influx of Atlantic water
10.0–7.8	Cibicides wuellerstorfi Globigerina quinqueloba	Increased oxygenation. A 2–3°C increase in temperature in <100 years c. 9.9K years BP. Increased warm Atlantic water
11.0–10.0	Reduction in abundance of foraminifera	Younger Dryas cooling
12.5–11.0	Cassidulina teretis Neogloboquadrina pachyderma Oridosalis umbonatus	Increased productivity and biodiversity. Salty, cold, bottom water. Reduced sea-ice cover. Increase of surface-water temperature from −1°C to 5.5°C
Pre-12.5	Elphidium excavatum Cassidulina reniforme	Oxygen-rich waters. Influx of warm Atlantic water. Near-glacial shallow-marine environments. Low producivity. Ice-covered surface water

Source: (based on Hald and Aspeli, 1997) See Figure 7.1 for location.

reduced sedimentation and foraminiferal production; at its close, surface-water temperatures rose rapidly to reach values similar to those of today in less than a hundred years (Hald and Aspeli, 1997). The environmental changes that were unfolding on land during this late glacial period have been reflected in pollen assemblages from numerous localities in Norway, Sweden and Finland. Synthesis of such work have been presented by Moe *et al.* (1996), Berglund *et al.* (1996) and Vasari (1996) respectively, along with syntheses of Holocene palaeoenvironmental research. The latitudinal spread of such countries makes generalisation difficult but it is evident that the late glacial period was one of considerable degree and rapidity of environmental change.

There is abundant evidence for environmental change during the Holocene in Scandinavia. Much of this is in the form of pollen diagrams (see references above). The chief development was the establishment of a boreal-forest type vegetation similar to that which characterised the last (Eemian) interglacial (see above). The character of the climatic climax boreal forest was not uniform, mainly due to the influence of latitude and altitude. In general, by *c.* 5K years BP, birch and pine dominated northerly latitudes and high altitudes, whilst at more southerly latitudes and lower altitudes, deciduous species, including alder, elm, oak and hazel were present. The later Holocene witnessed the arrival and expansion of spruce and, in some regions, beech and hornbeam.

According to Donner (1995), there was renewed glacial activity in the Scandinavian Mountains at *c.* 2.5K years BP. The formation of new glaciers in areas free of glacial ice since *c.* 8K years BP reflects a period of climatic deterioration. However, these glaciers advanced again, this time to their lowest elevations, by 1700–1750 AD. This was the time of the Little Ice Age (Chapter 6), the evidence for which comprises glacial moraines and documentary sources that detail the destruction and abandonment of farms (Lamb, 1995).

7.4 Eurasia

This extensive region extends from the Atlantic Ocean west of Norway, to the Pacific Ocean east of Siberia. As reference has already been made to the Scandinavian sector of Eurasia (Section 7.3), this section will examine the environmental history of Eurasia east of Finland.

The evidence from marine sediments in the Arctic (Section 7.2) attests to the repeated cold/warm stages that influenced this region over the last 3×10^6 years. Consequently, the vast tracts of boreal forests and peatlands that characterise the region today are the culmination of relatively recent environmental change since the opening of the Holocene about 10K years ago.

According to Arkhipov *et al.* (1986a, b) there is evidence in Siberia and North-east

Eurasia (in the Kamchatka region) for three main glaciations. These are given in Table 7.3. In Siberia, each of the three periods of glaciation was characterised by two glacial advances separated by an interstadial period. The earliest of the glacial stages, the Shaitan in Siberia, is considered to be equivalent to the marine oxygen isotope stage 12 (Figure 2.4). This means that, although there is little evidence for it, this vast area experienced many earlier cold/warm periods. There is also uncertainty about the extent of ice during these glacial stages. For example, Dawson (1992) in his synthesis of the events that occurred during the last ice age, indicates that much of western Siberia and North-west Eurasia may have been ice free, with the major centre of ice accumulation being over Franz Josef Land and Novaya Zemlya (Figure

Table 7.3 *The late Quaternary stratigraphy of Siberia in relation to that of Northern Europe*

	Northern Europe	Siberia	Siberian main events
G	Weichselian	Zyrianka	Sartan glacial advance
			Karginsky interstadial
			Early Sartan (Zyrianka) glacial advance
IG	Eemian	Kazantsevo	
G	Saale Complex	Bakhta	Taz glacial advance
			Shirta interstadial
			Samarovo glacial advance
IG	Holstein	Tobol	
G	Elster Complex	Shaitan	Late Shaitan glacial advance
			Unnamed interstadial
			Early Shaitan glacial advance
IG	Cromer	Talagaikino	

G = glacial stage.
IG = interglacial stage.

Source: Based on Arkhipov *et al.* (1986a, 1986b).

7.1). Where the ice was absent, periglaciation occurred; certainly it occurred above latitude 60°N, and probably extended as far south as latitude 50°N.

Evidence for the last interglacial period (Kazantsevo) confirms evidence from numerous palaeoenvironmental archives for interglacial conditions that were warmer than those of today. Lozhkin and Anderson (1995) have examined alluvial, fluvial and organic deposits in river terraces in north-east Siberia and have shown that larch (*Larix dahurica*) forests extended north-west by 600 km, along with a north-westward extension of birch (*Betula ermani*), as compared with their present location. This indicates that average July temperatures were between 4°C and 8°C warmer than at present. There is also evidence for a substantial marine transgression during the last interglacial, when relative sea-levels were 10–15 m higher than today (Alekseev, 1997). As the subsequent glaciation caused ice accumulation at the poles, the sea-level declined.

Research on the sediments of Lake Lama in Central Siberia (Hahne and Melles,

1997) provides evidence for environmental change during the last 17K years. The fact that sedimentation and pollen accumulation were occurring in the lake basin at this time, confirms that this region was not glaciated. Moreover, the late glacial oscillations recorded in the numerous archives referred to in Chapters 3, 4 and 5 are evident in the sediments of Lake Lama. A summary of the environmental changes that occurred during this climatically dynamic period is given in Table 7.4. The pollen analysis of the site shows that dense larch forests developed in the region in the early Holocene, with larch–spruce forests occurring by the mid-Holocene. Some oscillation of climate is indicated by the fluctuating proportions of spruce pollen, but the last 1000 years has witnessed climatic deterioration, causing the replacement of forest vegetation by tundra. Similar research on the sediments of Lake Elgennya, of the Upper Kolyma region in north-eastern Siberia (1040 m above sea-level), which extend back to 15K years BP, also testify to the absence of ice during full-glacial time when there was a herb-willow (*Salix*) tundra vegetation cover (Anderson *et al.*, 1997). As the Holocene opened, the shrub-tundra community altered in character to include birch and alder, possibly with the presence of some coniferous forest which was dominated by larch at *c.* 9.4K years BP. By 8.6K years BP, the shrub *Pinus pumila* had appeared and the vegetation community was similar to that of the present day. These two studies reflect regional and altitudinal differences in climatic and vegetation change.

7.5 North America

Whilst ocean-sediment cores from the North Atlantic and North Pacific attest to repeated cold and warm stages during the last 3×10^6 years in this region, there are various deposits on the continent north of 60° latitude that relate to many of the cold stages. These occur in Arctic Canada and Alaska, though the older the deposits are, the more difficult it becomes to relate them to each other and to marine oxygen isotope stratigraphy. Such difficulties arise because of the fragmented nature of the records and the inadequacies of age-determination techniques.

In Alaska, for example, Hamilton (1986) has presented evidence for Pliocene glaciations i.e. before 1.8×10^6 years BP. This evidence includes erratics from the Gunsight Mountain; these and other glacial deposits such as those of Skull Creek and Iron Creek in the Seward Peninsula (Figure 7.1) may have been deposited *c.* 2.8×10^6 years BP. Moreover, Hamilton states: 'The mountains of southern Alaska probably were glaciated repeatedly in late Tertiary time'. Similarly, there is fragmentary but widespread evidence, mainly in the form of glacial drift, for repeated glaciation during the early and middle part of the Quaternary period. In Arctic Canada, there is evidence for Pliocene and early Quaternary glaciations, especially on Banks Island, Baffin Island and in the Hudson Bay region. The repeated advance of the ice and its demise has contributed substantially to the complexity of landform genesis in these areas.

Table 7.4 *A summary of the environmental changes recorded in the sediments of Lake Lama, south-western Taymyr Peninsula, central Siberia, over the last 17K years*

Age (K years BP)	Period	Characteristics
H 0 — Present	Sub-Atlantic	Some **climatic oscillation**: cool, warm, cool. Dense spruce–larch forest opens up a little and non-arboreal taxa increase
O 2.5	Sub-Boreal	Some **climatic oscillation**: cool, warm, cool. This is reflected in oscillations of spruce. Spruce–larch forest dominates the landscape
L 5.0	Atlantic	**Climatic optimum**. The vegetation was dominated by spruce, probably with some larch. There may have been climatic oscillations
O 8.0	Boreal	Continued **climatic warming**, but with increased rainfall. The latter possibly caused larch to decline, while shrub alder and spruce (*Picea obovata*) increased
C 9.2	Preboreal	**Climatic warming**. Shrub birch, willow and juniper dominated initially. Later, tree birch, larch and poplar became established
E N E 10.3	Younger Dryas	**Cold, possibly moist climate**. Decline of dwarf birch and no evidence of larch. Sedges (Cyperaceae) were important, along with other non–arboreal species
L A T 11.0	Allerød	**Climatic warming**. Spread of dwarf birch and decline of species indicative of arctic conditions. Larch may have been present
E G L 12.0	Older Dryas	**Cold, dry climate**. Similar vegetation to the Oldest Dryas
A 12.3	Bølling	**Climatic warming**. Spread of dwarf birch and decline of species indicative of arctic conditions
C I 13	Oldest Dryas	**Cold, dry climate**. Species indicative of an arctic environment predominate, e.g. *Artemisia*
A L 17		

Source: Based on Hahne and Melles (1997)

The same problem of age estimation and correlation pertains to deposits of similar age in the Canadian Northwest Territories. However, recent research by Duk-Rodkin *et al.* (1996) has revealed a stratigraphic sequence in the Canyon Ranges of the Mackenzie Mountains (Table 7.5), which provides evidence for the reconstruction of environmental change since the Pliocene. The presence of numerous glacial deposits separated by palaeosols provides evidence of repeated glacial–interglacial cycles. A summary is given in Table 7.5. The characteristics of the glacial deposits indicate that the ice originated locally within the Mackenzie Range and was thus of cordilleran, i.e. mountain, rather than continental origin. The evidence also indicates that the only continental ice sheet emanating from the Arctic zone was that of the late Wisconsin Laurentide ice sheet *c.* 30K years BP, i.e. during a **stadial** occupying the latter part of the last ice age.

Prior to the last major continental ice sheet, the Wisconsin in North America, which is equivalent to the Weichsel in Northern Europe (Fig. 2.4), there was an **interglacial stage**. In North America this is known as the Sangamon (it is equivalent to the Eemian in Northern Europe; **oxygen isotope stage** 5e). Evidence from numerous and diverse sources indicates that this interglacial was, at its optimum, warmer than today. This premise is substantiated by pollen analysis from Banks Island (Arctic Canada, see Figure 7.1 for location), where the last interglacial is referred to as the Cape Collinson interglacial (Andrews *et al.*, 1986). During this period, sea-level was at a high, inundating continental-shelf areas and severing land bridges between what are today islands of the Canadian Arctic, and between Alaska and north-east Asia. In general, however, little is known in detail about the environment of the Sangamon stage in North America north of latitude 60°N. Evidence of this period (*c.* 130K to 120K years BP) from Greenland ice cores confirms the magnitude of climatic amelioration when compared with the pre- and post-glacial stages (Chapter 4).

The last glacial stage in northern North America was dominated by the Laurentide ice sheet, as shown in Figure 5.3(A). This had two centres of accumulation: the Keewatin and Labrador centres, and the ice sheet reached its greatest extent between 22K years and 17K years BP in parallel with the ice sheet that covered much of Northern Europe. In North America there was also a cordilleran ice sheet emanating from the Rocky Mountains; this reached its maximum extent *c.* 15K years to 14K years BP, though it is not obvious why the expansions of the two ice caps were asynchronous. As Figure 5.3(A) shows, ice covered most of North America above 60°N; only parts of Alaska were ice-free, and these were characterised by a tundra environment. Moreover, the lowered sea-level of this period meant that there was a land bridge between Alaska and north-east Asia. This facilitated the migration of plants and animals into North America as the ice retreated, and may have provided a route for people to enter the Americas, especially as an ice-free corridor developed *c.* 12K years BP between the Cordilleran and Laurentide ice sheets. This, however, is a controversial

Table 7.5 *A summary of environmental change over the last 2.5 × 10⁶ years in the Mackenzie Mountains, Northwest Territories, Canada*

Unit	Palaeoenvironmental inferences	Age (K years BP)
	Modern soil	
Mackenzie Lowland Formation	The glacial deposits derive from the Laurentide (continental) ice sheet rather than from montane sources	30–22
Brunisol (Palaeosol 5)	Brown soil	
Loreta Formation	The most recent montane glacial deposit, which accumulated during oxygen isotope stage 8, 12 or 16	c. 200
Brunisol (Palaeosol 4)	Thin brown soil which developed under conditions similar to those of today	Age uncertain
Little Keele Formation	Glacial deposits	
Eutric Brunisol (Palaeosol 3)	Brown soil development	Age uncertain but is probably <780
Rouge Mountain Formation	Glacial and periglacial features evident in several localities	
Luvisol 2 (Palaeosol)	This is a well-developed palaeosol reflecting a long, warm period	c. 1950
Abraham Formation	Glacial till deposited during oxygen isotope stage 70, 66 or 64. Evidence for more than one cold stage, e.g. frost features	Olduvai 1770–1950
Luvisol 1 (Palaeosol)	This developed on the Inlin Brook till, under conditions warmer and drier than today	c. 2000
Inlin Brook Formation	Earliest till deposited between oxygen isotope stages 100–72	2580–2000
Basal colluvium	Pediment formation	Miocene–Pliocene unconformity

Source: Based on Duk-Rodkin *et al.* (1996).

subject, as there is disputed evidence for the presence of humans in the Americas as early as 30K years BP.

According to Evison *et al.* (1996), there is evidence for renewed glacial advance in the Brooks Ranges of Alaska, during the Little Ice Age in about the 1600s and for glacier retreat during the twentieth century. The latter has occurred at average rates of between 2 m and 19 m annually since 1890. In addition, numerous studies on lake sediments and peat deposits provide evidence for the re-establishment of vegetation and animal communities on a deglaciated landscape. One example of such research is that of Szeicz *et al.* (1995) on the late glacial and Holocene environmental history of the central Mackenzie Mountains of the Northwest Territories. Investigations involving the sedimentary and pollen record of three lakes in the tundra, forest-tundra and open forest zone provide evidence of environmental change during the last 12K years. By this time, the Laurentide ice sheet was beginning to retreat. Herb-tundra vegetation communities colonised the deglaciated landscape, and were replaced by shrub-tundra, in which dwarf birch (*Betula glandulosa*) predominated, by 10.2K years BP. Balsam poplar (*Populus balsamifera*) had also migrated into the region by this time and may have extended beyond the present tree-line. By 8.5K years BP, spruce (*Picea* spp.) populations were expanding; by 6K years BP, vegetation communities included alder (*Alnus*), as is the case today.

At some sites in North America above 60°N, there is evidence for a climatic reversal equivalent to the **Younger Dryas** in Europe (Section 3.2.1). On Pleasant Island off the coast of south-eastern Alaska, for example, the immigration of lodgepole pine (*Pinus contorta*) and mountain hemlock (*Tsuga mertensiana*), which began at *c.* 13K years BP, was interrupted between 10.6K years BP and 9.9K years BP as a return to colder conditions was characterised by an expansion of tundra species (Hansen and Engstrom, 1996). During the early Holocene, western hemlock (*Tsuga heterophylla*), and mountain hemlock, replaced sitka spruce (*Picea sitchensis*), and by 7K years BP tracts of open muskeg (i.e. swampland) had developed. This provided favourable conditions for the renewed immigration of lodgepole pine, but by 3.4K years BP the vegetation communities were dominated by bog species. This may reflect the development of a climate cooler and wetter than that characteristic of the earlier Holocene.

These and many similar studies illustrate some overall patterns in the nature and direction of environmental change in this region of North America. However, such studies also highlight the significance of local conditions such as topography and altitude as well as proximity to the source areas and migration routes of invading plant species, as determinants of the characteristics of Holocene vegetation communities.

7.6 The Southern Hemisphere south of latitude 60°S: Antarctica

The only landmass south of latitude 60°S is the continent of Antarctica. Until recently, it was generally considered that the Antarctic ice sheet had grown to its present size by $c.$ 14×10^6 years ago. However, Barrett, in his review of Antarctica's geological history (Barrett, 1991), has pointed out that deposits containing diatoms (Figure 3.4) have been discovered in the Transantarctic mountains and have been radiometrically dated at $c.$ 3×10^6 years BP. Such materials could only have been deposited in warmer times when sea-levels were high; the sediments were subsequently transported by ice to their present location. In addition, lake sediments with remains of the southern beech (*Nothofagus*) have been found in these mountains, again indicating the possibility of warm conditions and even the possibility of a forest cover (Andersen and Borns, 1994). This evidence, along with additional data from marine sediment cores extracted from the Southern Ocean and further marine deposits found in Antarctica itself, indicates that warming occurred during the Pliocene. The deposits indicate that this must have been substantial, causing a considerable sea-level rise. Thereafter Barrett suggests that Antarctica developed into its present condition, with an ice sheet as extensive as it is today. Such a possibility is, however, refuted by Hall *et al.* (1997), whose recent work on glacial deposits in the Transantarctic Mountains indicates that the East Antarctic ice sheet has been stable with little melting for at least 3.8×10^6 years. This aspect of Antarctica's environmental history thus remains controversial.

It appears that little is known about the impact on Antarctica of the many cycles of warming and cooling that have occurred during the last 3×10^6 years. This is partly because evidence is buried within and between vast areas of deep ice, or is located beneath the ocean of the continental shelf, and partly because of the enormous difficulties of undertaking research in such a remote and climatically extreme environment. However, the ice cap is itself an archive of palaeoenvironmental data, as has been discussed in Chapter 4, and the oxygen isotope record in particular attests to the oscillation of warm and cold periods over the last $c.$ 160K years, and the rapid shift from cold to warm conditions as the penultimate and last ice ages drew to a close. During the ice ages, the surface temperatures were much lower than today, by at least 8°C, and precipitation was much reduced. However, much of the data collated from Antarctic archives provide a greater insight into global or Southern Hemisphere environmental change than they do into local Antarctic environmental change.

Inevitably, and as is the case elsewhere, the vast majority of information is available about environmental change during the last glacial and the current interglacial periods. During the last ice age, as compared with the present, sea-ice, for example, extended well beyond the Antarctic Circle. This was probably the case during earlier ice ages though there is considerable dispute about the increase in ice-sheet thickness and sea-ice extent during the last ice age (see discussion in

Williams *et al.*, 1993). There is also a dispute about the timing of deglaciation, which may have been later in Antarctica than in the Arctic. For example, in the Arctic, deglaciation began *c.* 11K years to 10K years BP, whereas in the Antarctic deglaciation may have occurred as late as *c.* 7K years BP. Certainly, geo-morphological evidence from James Ross Island in the Antarctic Peninsula indicates that the initial deglaciation of areas that are currently ice free occurred *c.* 7.4K years BP, and was associated with a 30 m rise above the present sea-level (Hjort *et al.*, 1997). There was a glacier re-advance, which extended into the ocean, by 4.6K years BP, and a recession by 4.3K years BP. In contrast, Licht *et al.* (1996) report that deglaciation began *c.* 11.5K years BP in the Ross Sea, and that the Ross Ice Shelf reached its present position *c.* 7K years BP. Today, there is much concern about the stability of parts of the Antarctic ice sheet and the impact that melting might have on sea-level and global climate. The West Antarctic ice sheet is considered particularly vulnerable to melting attributable to possible global warming.

7.7 Conclusions

High latitudes today provide considerable evidence for environmental change in the past. The polar regions, for example, provide not only archives of palaeoenvironmental change within the ice caps themselves (Chapter 4) but also modern analogues for past processes. In the Northern Hemisphere, all the extensive land area above latitude 60°N must have experienced direct glaciation at some time in the past 3×10^6 years. The characteristics of glacial deposits facilitate the identification of source areas, the direction of ice movement, and its overall extent. Although the record is fragmentary and thus difficult to interpret, it must be concluded that no two glacial stages, or two interglacial stages, were identical. Moreover, it is apparent that both continental ice and cordilleran ice played major roles in shaping landscapes above latitude 60°N, especially in North America. Here the last ice sheet was particularly extensive, reaching farther south than the Great Lakes. In contrast, much of Eurasia was ice free and was characterised by arctic tundra vegetation communities.

As climate ameliorated and the ice retreated with the onset of the Holocene, newly deglaciated landscapes throughout the northern landmasses were invaded by tundra species, and in many regions boreal forest had developed by *c.* 5K years BP. The dominant species varied according to continent, altitude and local conditions. Evidence of environmental change during this period is abundant, and includes a wealth of information available from lake sediments, notably pollen assemblages.

For the Southern Hemisphere south of 60°S, considerably less is understood about the environmental changes of the last 3×10^6 years. The reaction of the

Antarctic ice sheet to the repeated cycles of warming and cooling remains to be determined. During the last ice age, the continental ice sheet and sea-ice were more extensive than they are now. It is also possible that the deglaciation of those parts of Antarctica that are currently ice free began later than similar events in the Northern Hemisphere.

Summary Points

- High-latitude regions experienced repeated **glacial**/periglacial conditions which alternated with **interglacial** conditions over the last 3×10^6 years.

- In terrestrial environments, the record of environmental change is fragmentary, and little is known about early glacial/interglacial cycles.

- In North America there is a long record of environmental change; it reflects the impact of both continental and cordilleran ice.

- There is evidence from numerous archives that the last interglacial period was warmer than the present interglacial by $c.$ 1°C to 2°C (average July temperatures).

- There is abundant evidence for the last ice advance (Weichselian–Wisconsin) in Europe and North America.

- Much of Eurasia experienced periglacial rather than glacial conditions during the last ice age, and was characterised by arctic-tundra vegetation communities.

- The abrupt shift from glacial/periglacial conditions to interglacial conditions as the last ice age ended is widely recorded.

- The record of environmental change in high latitudes of the Southern Hemisphere is less well understood than that of the Northern Hemisphere; ice cores from the Antarctic ice cap provide direct evidence of changing conditions, but the geological/sedimentological evidence is often controversial.

General further reading

Ice Age Earth. A.G. Dawson. 1992. Routledge, London.

Late Quaternary Environments of the Soviet Union. A.A. Velichko (ed.). 1984. University of Minnesota Press, Minneapolis.

Late Quaternary Environments of the United States, Vol. 1. The Late Pleistocene. H.E. Wright (ed.). 1983. University of Minnesota Press, Minneapolis.

The Ice Age World. B.G. Andersen and H.W. Borns Jr 1994. Scandinavian University Press, Oslo, Copenhagen and Stockholm.

The Quaternary History of Scandinavia. J. Donner. 1995. Cambridge University Press, Cambridge.

8 Environmental change in middle latitudes (latitudes 30–60°N and 30–60°S)

8.1 Introduction

Figure 8.1 illustrates the areas of the Earth's surface with which this chapter is concerned. In terms of a global perspective, the middle latitudes are characterised by temperate, Mediterranean-type and steppe vegetation communities and climates. The greater extent of land area in the Northern Hemisphere than in the Southern Hemisphere means that the terrestrial record of environmental change is most extensive in the former. There are, however, numerous ocean-sediment

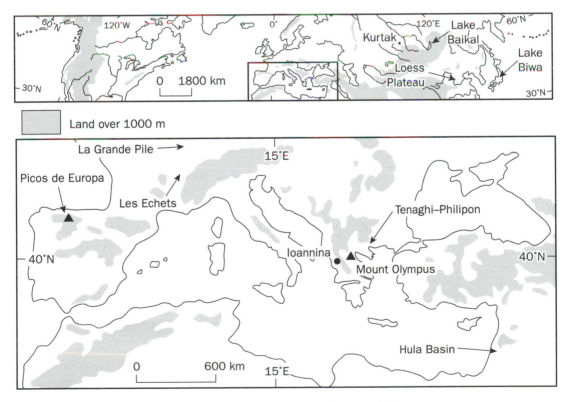

Figure 8.1 *Middle latitudes: the location of sites and regions referred to in the text*

cores from both hemispheres, which provide an unbroken record of environmental change for much of the last 3×10^6 years.

In both hemispheres not all of the land surface directly experienced the repeated ice ages of this period. Those lands that were not covered with ice but that were adjacent to glaciated areas experienced periglaciation. The remaining lands of the middle latitudes experienced environmental change, driven by the temperature changes associated with the repeated oscillations of warm and cold stages. Such oscillations influenced soil processes as well as the composition and dynamics of ecosystems. Evidence for the nature and direction of environmental change derives from a variety of sources. For example, there are several long lake sequences, notably in Italy and Greece, and an increasing body of evidence from Lake Baikal in the Russian Federation. Many of the techniques of palaeoenvironmental investigation discussed in Chapter 5, especially pollen analysis, have been applied in a wide variety of contexts to provide evidence of both natural and cultural environmental change, as illustrated by many of the case studies discussed in the following sections. There is also an increasing amount of information emanating from the middle latitudes of the Southern Hemisphere. Together with long records from ocean sediments, information from Australia, New Zealand, South Africa and South America is contributing to the understanding of global environmental change, as well as providing an insight into local and regional environmental change.

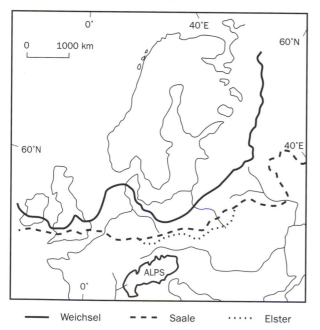

Weichsel — — — Saale ····· Elster

Figure 8.2 *The maximum extent of the last three major ice advances in Europe*

Source: Based on Andersen and Borns (1994).

The wealth of information on natural environmental change in middle latitudes is now considerable. What follows therefore is, of necessity, a basic introduction.

8.2 Europe

Figure 8.2 illustrates the southernmost limits of the major ice advances during the last three ice ages. Apart from regional centres of ice accumulation in, for example, the Alps, most of Europe south of latitude 50°N did not experience direct glaciation. Consequently, between latitudes 50°N and 60°N there is geomorphological evidence on which to reconstruct the maximum extent of ice sheets during the last three ice ages

(Figure 8.2). In Europe south of these ice limits, there is a wide range of evidence to facilitate the reconstruction of environmental change in areas not directly affected by the presence of ice sheets. One of the most complete sequences of sediments is that of the Rhine Delta, with a record covering $c.\ 2.5 \times 10^6$ years. Although there are several other sites with a 'long' record of environmental change, such as the deep peat and lacustrine sequences of Greece, Italy (Section 5.3.1) and France, none represent more than 1×10^6 years. However, such records are invaluable for establishing stratigraphic links with the marine-sediment record, notably the oxygen isotope record (Chapter 3); they thus contribute to the establishment of a global framework for environmental change during the last 3×10^6 years.

The composite stratigraphy of The Netherlands is given in Figure 8.3 (see also Figure 2.4). This reflects Zagwijn's (1992) view that the Pliocene/Pleistocene boundary, i.e. the onset of the Quaternary period, should be placed at 2.3×10^6 years (Section 2.3). The stratigraphy reflects the oscillation of warm/cold stages, in common with the records from other archives. However, a distinction is drawn between the **interglacial** stages older than 900K years and those younger than 900K years. The pollen assemblages of the older interglacials do not show a clear succession of forest taxa, as most of the taxa appear to have been present, but with reduced populations, throughout each warm stage, rather than showing a pattern of immigration, spread and decline. This may reflect the fact that these early interglacials experienced relatively high temperatures within a weak cold–warm cycle, as compared with the later interglacials, which were part of cold/warm cycles characterised by greater temperature differences. Prior to 900K years BP, the climatic cycles were of a shorter duration (40K years) than those after 900K years BP (120K years). Zagwijn suggests that these later interglacials experienced either an oceanic climate associated with high sea-levels or a continental climate associated with lowered sea-levels.

Evidence for intensified cooling at $c.\ 2.3 \times 10^6$ years is available elsewhere in Europe. For example, a marine regression and intensified cooling are indicated by sedimentary and fossil evidence from the southern part of the North Sea Basin (Funnell, 1996). As discussed in Section 3.2.1, there is evidence for ice rafting around this time in the North Atlantic Ocean. Clearly, this was a significant time horizon in Europe's environmental history. Nevertheless, in areas directly affected by the formation and advance of ice sheets, there are no archives that represent these early events. For example, in the Alpine region, there is little unequivocal evidence for glaciation prior to $c.\ 2 \times 10^6$ years (Schlüchter, 1986; Billard and Orombelli, 1986). There is, however, abundant evidence for several later glaciations. Indeed it was the deposits left by these ice advances that inspired the original stratigraphic framework for the Alpine region, which was formulated by the famous professors of Geography, Albrecht Penck and Edward Brückner (Section 1.2.1; Figure 1.2; Nilsson, 1983). They proposed that there had been four major glaciations, with intervening interglacials. The original terminology is still

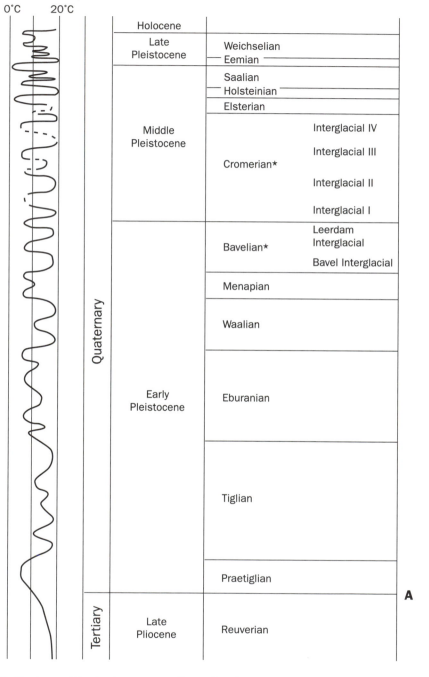

Estimated mean temperature in July

A   The base of the Quaternary according to Zagwijn; *Several glacials and interglacials within one period

Figure 8.3 *The stratigraphy of the last 2.5 × 10⁶ years in The Netherlands*

Source: Based on Zagwijn (1992).

used today and is given in Figure 2.4. The extent of the Alpine ice caps during the last (Würm) glaciation and earlier two (Riss and Mindel) glaciations is given in Box 8.1.

In other upland areas of Europe's middle latitudes, there is also evidence for repeated glaciation. For example, in the mountains of the north-west Picos de Europa of northern Spain (see Figure 8.1 for location) there have been at least five periods of glaciation, the most extensive of which is considered to be of early Quaternary age, i.e. it is no younger than 1.8×10^6 years (Gale and Hoare, 1997). The record of the occurrence of glaciation is not so long on Mount Olympus, Greece. Nevertheless, Smith *et al.* (1997) report that a series of sedimentary sequences comprise glacial deposits, periglacial deposits and soils. A summary of inferences based on the sedimentary sequences is given in Table 8.1. In general, Smith *et al.* suggest that three glacial episodes occurred, though they do not rule out the possibility of earlier **glacial** episodes whose deposits may have been reworked.

In relation to continental sequences, however, there are none that provide an unbroken record of environmental change over the last 3×10^6 years. Amongst the longest records are the peat sequence of Tenaghi–Philippon and the sediments of Ioannina, Greece (Section 5.3.1). The former represents accumulation during the last 1×10^6 years. Both reflect the oscillation of warm and cold stages, as shown in Figure 8.4. This gives a possible correlation between the two, based on changes in arboreal (tree) pollen/non-arboreal (shrubs, herbs) pollen percentages. In the regions of Greece of which these sequences are representative, namely Macedonia in the case of Tenaghi–Philippon, and the Pinus Mountains of the north-west in the case of Ioannina, cold stages were characterised by open vegetation communities dominated by grasses and herbs, whilst warm stages were characterised by forests dominated by pine (*Pinus*) and oak (*Quercus*). However, populations of trees were maintained even during the cold stages, as is especially apparent at Ioannina. Tzedakis suggests that the region around Ioannina may

Table 8.1 *The glacial/interglacial Quaternary history of Mt Olympus, Greece*

Oxygen isotope stage	Event	Environmental processes
1	Holocene	Stream incision; soil formation
2, 3, 4	Glaciation 3	Glaciation at valley heads only; glaciers extended during stage 2/3
5	Interglacial	Erosion, soil formation
6	Glaciation 2	Upland ice and limited valley glaciers
7	Interglacial	Erosion followed by **pedogenesis**, i.e. soil formation
8	Glaciation 1	Upland ice, and valley glaciers which extended as piedmont lobes to the east, north and west of Mt Olympus

Source: Based on Smith, (1997), with modifications.

Box 8.1

The glaciations of the Alps

1 Penck and Brückner, pioneers of glacial stratigraphy and chronology in the late nineteenth and early twentieth centuries (see Section 1.2.1 and Figure 1.2), suggested that the Alps had experienced four major glacial stages.

2 These are, in order of increasing age, the Würm, Riss, Mindel and Gunz (Figure 1.2).

3 Evidence for the Würm is most extensive; during this period the snow-line was *c.* 1200 m lower than at present.

4 Evidence for the earliest glacial stages is less abundant, but end moraines indicate that the snow-line was a further 100–200 m lower than at present.

5 The map below illustrates the limits of the Würm, Riss and Mindel ice advances.

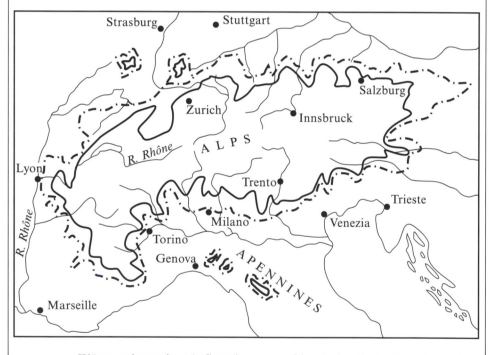

——— Würm end moraines (reflect the extent of ice during the last ice advance)
— · — · Riss and Mindel end moraines (reflect the extent of ice during earlier ice advances)

Source: Based on Nilsson (1983)

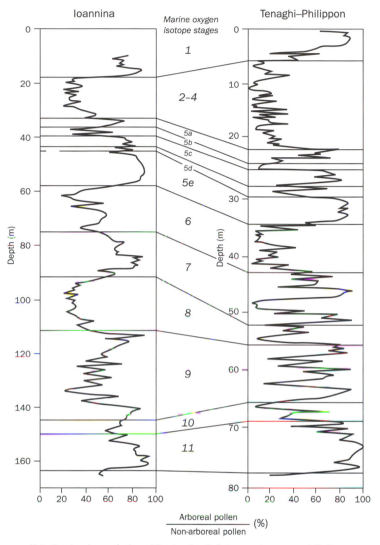

Figure 8.4 *The relationship between the Tenaghi–Philippon and Ioannina (Greece) pollen sequences*

Source: Based on Tzedakis (1993).

Note the abundance of arboreal (tree) pollen during warm stages, especially the interglacials, e.g. stages 1, 5e.

have been a 'long-term' **refugium**, from where tree populations migrated when the climate ameliorated.

The last glacial/interglacial cycle (i.e. 140K to 10K years BP) is well represented in the mires of La Grande Pile in the Vosges and Les Echets near Lyon, France. According to Lowe and Walker's (1997) synthesis of data from these sites, the pollen assemblages indicate that the last interglacial stage was warmer than the present interglacial (the Holocene), as is indicated by other types of evidence, and that there were at least two interstadials during the last ice age. Also, in common with data from many other archives, the last glacial period ended relatively abruptly, with rapid climatic/ecological changes characterising the period known as the late glacial which is transitional to the Holocene. This period has been discussed in Section 4.5, and because of the rapidity of climatic change, it has been the focus of considerable research interest. A synthesis of environmental change in Ireland, Britain, Belgium, The Netherlands and north-west Germany has been presented by Walker *et al.* (1994), a summary of which is given in Figure 8.5. As in many palaeoenvironmental archives, the temperature regression of the **Younger Dryas** is especially notable; in upland regions the temperatures declined sufficiently to promote glacial readvances. Details of the late glacial environment in Britain and Greenland are discussed in Box 8.2.

Box 8.2

The late glacial period in Britain and Greenland

1 The late glacial is the transitional period between the last ice age and the Holocene; it occurred between 14K and 10K years BP.
2 It was complex in terms of climatic and ecological change.
3 It is registered in a variety of archives, and is the focus of considerable research because of the rapidity with which environmental change occurred.
4 Numerous sites in Britain have yielded information on this particularly dynamic period.
5 Pollen and coleopteran (beetle) analysis, along with radiocarbon age determination, have been used to reconstruct the environmental changes during this *c.* 4K year period. The results are illustrated below and show the rapid oscillation of cold and warm periods.
6 Temperature reconstructions based on coleopteran assemblages from Gransmoor, East Yorkshire, correlate with snow accumulation rates in the GISP2 ice-core (Greenland). High temperatures correlate with the greatest rates of snow accumulation, and are considered to reflect the broad synchroneity of climatic change in Greenland and Britain.

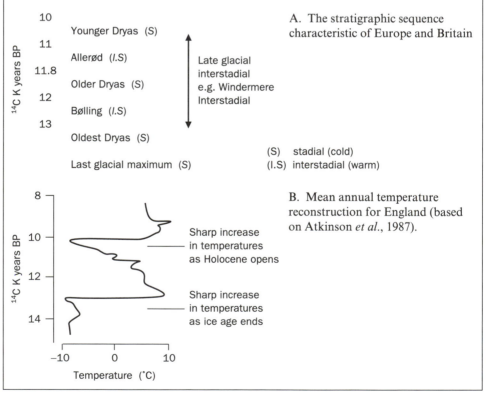

A. The stratigraphic sequence characteristic of Europe and Britain

(S)   stadial (cold)
(I.S) interstadial (warm)

B. Mean annual temperature reconstruction for England (based on Atkinson *et al.*, 1987).

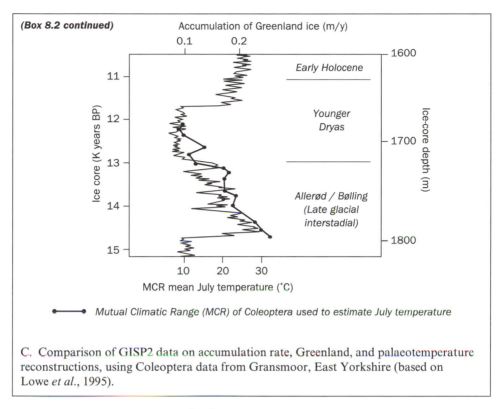

(Box 8.2 continued)

C. Comparison of GISP2 data on accumulation rate, Greenland, and palaeotemperature reconstructions, using Coleoptera data from Gransmoor, East Yorkshire (based on Lowe *et al.*, 1995).

Environmental change in much of Europe during the Holocene has been discussed elsewhere in this book. For example, British woodland history is examined in Sections 5.3.3 and 5.4, the dendro-ecological/dendroclimatological record for parts of Europe is discussed in Section 6.2.1, and the value of historical and meteorological records along with documentary evidence is considered in Sections 6.3.1 and 6.3.2. The latter records reflect climatic change during historical times.

8.3 Eurasia (including Japan)

As stated in Section 7.4 in relation to the high latitudes of Eurasia, the vast extent of the region makes a synthesis, and especially a brief synthesis, particularly difficult. The western part of Eurasia borders on the Baltic States and Balkan States and extends east to the North Pacific Ocean (Figure 8.1). Much of the region lay beyond the margins of the ice sheets of the last 3×10^6 years, especially east of longitude 40°E, though during the ice ages periglaciation would have been extensive (Figures 5.4; 8.1). This region also encompasses vast tracts of **loess**, which were deposited during successive ice ages (Section 5.6). The sedimentary sequences of the Loess Plateau of China, for example, provide a record of environmental change over the last 2.6×10^6 years (Figure 5.10). However, the

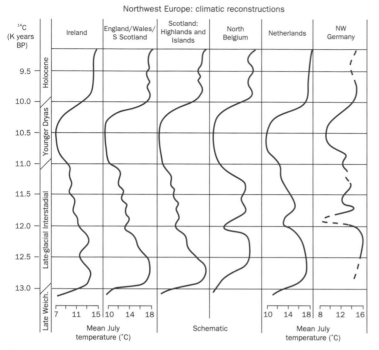

Northwest Europe: climatic reconstructions

Figure 8.5 *Summary of climatic reconstructions for the late glacial period in northwest Europe*

Source: From Walker *et al.* (1994).

sediments of Lake Biwa, in Japan, provide the longest and most complete continental record of environmental change in the world. Other long records include those of the Hula Basin, Israel (Section 5.3.1; Figure 5.6) and, possibly, Lake Baikal in Siberia, where investigations began in 1993 (see Figure 8.1 for locations).

The pollen assemblages of a 1422 m core from Lake Biwa, Japan, have been presented by Fuji (1988). The uppermost 900 m of this core represent a record of vegetation change during the last 3×10^6 years and the pollen record has been subdivided into 37 pollen zones. These show considerable variations in terms of pollen from the catchment, and include cyclical variations. During **glacial**/cold stages, a vegetation typical of subpolar regions characterised the mountainous area in the catchment of Lake Biwa. In the lowlands in the immediate vicinity of the lake a cool-temperature vegetation prevailed. In contrast, during **interglacial**/warm stages, and **interstadials**, temperature and cool-temperate communities occupied the mountains, whilst broad-leaved deciduous and evergreen trees of the temperate and warm-temperate zones predominated in the lowlands. As shown in Figure 5.6, there are parallels between the upper Biwa record and other long lacustrine sequences (Section 5.3.1).

Recently, Meyers and Takemura (1997) have examined sediment type and rates of organic matter accumulation in Lake Biwa. They report that, prior to 430K years BP, sediments were deposited in shallow-water and fluviodeltaic conditions within which occasional deposition in deep water occurred; after 430K years BP, deposition in deep-water conditions predominated. In terms of organic matter deposition, most originated from the algae that inhabited the lake waters. There is also a distinction between interglacials/warm stages and glacials/cold stages, insofar as the former were characterised by as much as nine times more organic matter accumulation. This is considered to represent the increased availability of nutrients leached from catchment soils and the rapid incarceration of organic matter within the sediments.

Initial coring of the Lake Baikal sediments has produced sediments spanning the last 500K years (Colman *et al.*, 1997). Preliminary palaeoenvironmental work, including the analyses of pollen and diatom assemblages, indicates changing conditions which paralleled the glacial/interglacial climatic changes. For example, pollen analyses indicate that during the last 250K years, vegetation communities comprising forest types and steppe types have alternated as the climate has oscillated between cold (glacial) and warm (interglacial) conditions. As research on Lake Baikal continues, it will generate valuable data that will contribute considerably to the elucidation of environmental change in Central Asia. Already relationships have been established with the deep loess sequences of China (Section 5.6) and with the marine oxygen isotope record (Chapter 3).

However, loess sequences occur in regions of Asia other than China. For example at Kurtak in southern Siberia some 1200 km west of Lake Baikal there is a 34 m loess sequence which contains a record of environmental change for the last *c.* 250K years, i.e. for the last two glacial/interglacial climatic cycles (oxygen isotope stages 1–7). Analyses of the mineralogy, grain morphology and magnetic characteristics have facilitated palaeoenvironmental reconstruction (Chlachula *et al.*, 1997). During the coldest periods, loess was deposited, this having originated from local granitic and metamorphic rocks; during periods of climatic amelioration, i.e. **interstadials**, loess deposition was interrupted and soil began to form. Interglacials were characterised by chernozem soils, i.e. black-earth soils that are prevalent in continental steppelands today. Extensive loess deposits further south, in the republics of Uzbekistan and Tajikistan for example, have also been investigated. Dodonov and Baiguzina (1995) have described a loess sequence from South Tajikistan, the deposition of which began between 2.5×10^6 and 2×10^6 years ago and thus corresponds with intensified cooling evident elsewhere (Section 2.3). Palaeosol characteristics from this sequence indicate that interglacial conditions were warmer and wetter than those of the glacial period.

Much further south, at latitude *c.* 33.5°N, the sediments of Israel's Hula Basin provide a record of environmental change during the last 3.5×10^6 years (Horowitz, 1989). In common with evidence from other archives, e.g. marine sediments and other long lake sequences (Chapter 3; Section 5.3.1), there is evidence for intensified cooling between 2.4×10^6 years and 2.6×10^6 years BP. The upper part of the Hula record is illustrated in Figure 5.6, which shows that forest was characteristic of the catchment during cold stages. Oak (*Quercus*) and pine (*Pinus*) were the dominant species, indicating that these cold stages were receiving higher levels of precipitation than the warm/interglacial stages. These latter were characterised by steppe/maquis (grassland/shrub) vegetation.

In relation to environmental change during the Holocene, including the historic period, examples have been given in Section 6.2.2 (tree-rings) and Section 6.3.1, e.g. Chinese drought–flood records.

8.4 North America

The Quaternary stratigraphy of North America is given in Figure 2.4 (Richmond and Fullerton, 1986). This shows that there were numerous ice ages/cold stages during the last 3×10^6 years, evidence for which derives from mountain (**cordilleran**) regions as well as from the central plains. However, no complete sequence has so far been discovered. As in Europe, there are problems with correlating between many disparate deposits, especially those of the late Tertiary/ early Quaternary, and with the marine oxygen isotope record. In contrast, the ice limits and dating of events during the Illinoian and Wisconsin cold stages are reasonably well documented. In the Great Lakes region, for example, deposits of the Illinoian, Sangamon and Wisconsin stages have been identified; events occurring within these stages have recently been re-evaluated by Johnson *et al.* (1997).

Beyond the directly glaciated area of North America, a record of environmental change can be obtained from a range of archives. For example, the packrat middens of the south-western arid zone have been referred to in Section 5.8, and the long lacustrine sequence of Owen's Lake, California, has been considered in Section 5.3.1 (see also Table 5.2). The sedimentary record of the Great Salt Lake in Utah and surrounding deposits have also provided palaeoenvironmental information. Today, the Great Salt Lake is much smaller and more saline than it has been in the past. This is typical of many lakes in the arid and semi-arid region of the south-western USA and, indeed, of arid and semi-arid regions elsewhere (Section 9.5). Lakes known to have undergone expansion and contraction in this way are referred to as pluvial lakes; their high-water stages and expansions are associated with periods of high rainfall. In contrast, reduced water levels and a contracted extent developed during the interpluvial conditions that often occurred during **glacial**/cold stages.

Figure 8.6 illustrates the former extent of the Great Salt Lake, i.e. the palaeo-lake Bonneville, *c*. 18K years ago when the last glacial maximum occurred. At this time Lake Bonneville was at its greatest extent; i.e. 51 300 km², which represented the combined extent of the present-day Great Salt Lake (i.e. 4000 km²), plus that of Sevier Lake and Utah Lake, and reached an altitude of *c*. 1600 m, which is 300 m higher than today (Benson and Thompson, 1987). Similarly, Lake Lahontan in Nevada occupied *c*. 22 300 km² at its greatest extent *c*. 13K years BP, in comparison with 1549 km² today (Benson and Thompson, 1987). According to Oviatt (1997), lake levels fluctuated on a millennial scale at the end of the last ice age, between 21K years BP and 10K years BP. Five major declines in water level, each of *c*. 50 m, occurred during this period.

Further records of environmental change are present in the extensive **loess** deposits of the USA (Figure 8.7), which are particularly deep in the major river valleys. One example is that of Barton County in Central Kansas (Feng, 1997),

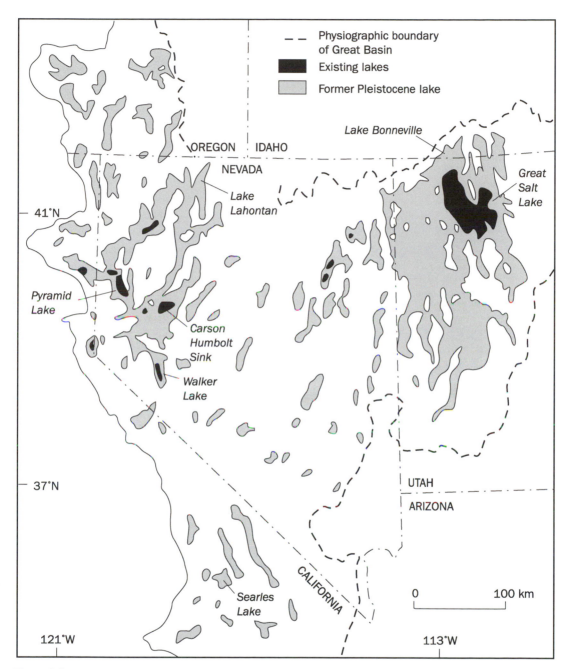

Figure 8.6 *The extent of lakes during the last ice age (Wisconsin) in the Western USA*

Source: Based on Benson and Thompson (1987).

Age (K years BP)	Barton County loess sequence	
	Unit	Characteristics
5	SURFACE SOIL	Silt with organic matter
	BIGNELL LOESS Warm, dry conditions; poorly weathered	Yellow-brown calcareous silt with calcium concentrations
8		
	BRADY SOIL Warm, moist conditions; strongly weathered	Grey-brown silt, with light brown-grey mottles and calcareous concentrations in lower levels
10		
	PEORIA LOESS Cold, dry conditions; little weathering	Yellowish to grey-tan silt with rusty-brown mottles and plant debris
20		
	GILMAN CANYON PEDOCOMPLEX Cool, moist conditions; strongly leached	Dark brown-grey silt with horizons rich in organic matter and a weakly structured B horizon
35		
	REDDISH PEDOCOMPLEX Mild climatic conditions; strongly weathered	Reddish silt, well-developed non-calcareous soil peds with clay and oxide coatings
70		
	BARTON SAND Fluctuating climatic conditions; moderate weathering	Weakly reddish, calcareous, well-sorted aeolian sand
92		
	First soil in Loveland Loess Warm interval; weathered First loess unit in Loveland Loess	Clayey and carbonate enriched with carbonate nodules and plant root canals Weakly reddish silt with powder carbonate
c. 193		
	Second soil in Loveland Loess Warm interval; weathered Second loess unit in Loveland Loess	Clay and carbonate enriched but fewer carbonate nodules than first soil unit or third soil unit Weakly reddish silt with powder carbonate
c. 260		
	Third soil in Loveland Loess Warm interval; weathered	Clayey, carbonate enriched abundant carbonate nodules

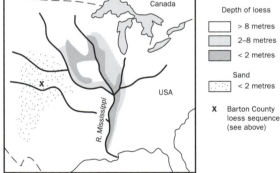

Depth of loess

☐ > 8 metres
▨ 2–8 metres
▣ < 2 metres

Sand
▢ < 2 metres

X Barton County
   loess sequence
   (see above)

Figure 8.7 *The extent of loess in the USA (based on Andersen and Borns, 1994) and the loess record in central Kansas (based on Feng, 1997)*

data from which are shown in Figure 8.7. The record extends back *c.* 260K years, and an analysis of the geochemical characteristics has provided a means of establishing the extent of weathering for each of the palaeosols present. As Figure 8.7 shows, the most weathered horizons correspond with periods of extended climatic amelioration whilst the least weathered sections correspond with periods of loess deposition during harsh climatic regimes.

Abundant records of environmental change during the late glacial period and the Holocene in the middle latitudes of North America are available. In relation to the late glacial period, recent research has led to the conclusion that the complex climatic/ environmental changes that characterised the 14K year BP to 10K year BP period in Europe (Figure 8.4) also influenced north-eastern North America. Some of the regional syntheses produced recently are given in Figure 8.8. Whilst there are variations between regions, the temporal differences in temperature are most pronounced for New Brunswick and least pronounced in southern New England. In all regions, the **Younger Dryas** cooling episode is recorded. In New England, for example, pollen assemblages of Younger Dryas age indicate that boreal tree species, notably spruce (*Picea*), larch (*Larix*), paper birch (*Betula papyrifera*) and alder (*Alnus*) increased at the expense of temperate species such as oak (*Quercus*) and white pine (*Pinus strobus*), which had colonised the area in the earlier, warmer period, i.e. the equivalent of the Bolling–Allerød (Peteet *et al.*, 1994).

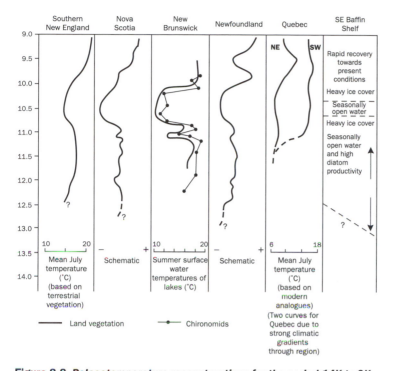

During the Holocene, patterns of vegetation communities similar to those of the present day developed as plant species migrated north to replace tundra communities, and temperate forests became established in New England and in the St Lawrence/Great Lakes region, for example. The extensive grasslands of the central plains, the prairies, established their present location as tree-lines and forests altered their locations, as shown in Box 8.3. Changes in fauna also occurred. Much of the detail of ecosystem changes during this period has been examined in Delcourt and Delcourt (1991).

Figure 8.8 *Palaeotemperature reconstructions for the period 14K to 9K years BP for southern New England (Peteet et al., 1994), Nova Scotia (Mott, 1994), New Brunswick (Cwynar et al., 1994), Newfoundland (Anderson and Macpherson, 1994), Quebec (Richard, 1994) and the south-east Baffin Shelf (Andrews, 1994)*

8.5 South America

The area of South America south of latitude 30°S is shown in Figure 8.9. This region has never experienced continental ice sheets similar to those of Europe or North America. However, the presence of the Andean Mountains means that glaciers and local ice caps developed, which altered landscapes through erosion and deposition. In areas unaffected by direct ice action, the depression of temperatures associated with the cold stages of the last 3×10^6 years prompted environmental change. Inevitably, little is known about the earliest **glacial/interglacial stages**, whilst an increasing amount of information about environmental change during the last 100K years, and especially the last 14K years, is beginning to provide a detailed picture of events and trends.

The oldest glacial deposits in the region are considered to be 3.6×10^6 years old (Andersen and Borns, 1994), a date produced from volcanic deposits interbedded with glacial deposits. According to Clapperton (1994), there is evidence for numerous Quaternary glaciations, but few dates are available to ascertain when

Box 8.3

Changes in the prairie/forest border in the Midwest of North America during the Holocene

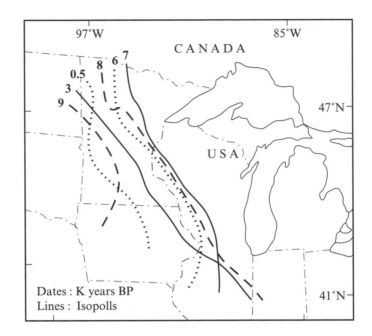

1 According to the pollen analyses of a range of lacustrine sequences in the Midwest of North America (Webb *et al.*, 1983) the prairie/forest border has shifted during the Holocene.

2 Pollen assemblages that represent prairie vegetation communities are characterised by *c.* 20 per cent prairie forb (herb) pollen. This includes ragweed (*Ambrosia*), sage (*Artemisia*), other Compositae and Chenopodiaceae/Amaranthaceae.

3 The map above shows the location of the 20 per cent isopoll (contours of pollen percentages) at different times during the Holocene.

4 At 9K years BP prairie first appeared; it reached its eastern limit *c.* 8K–7K years BP, with minor shifts between 8K years and 6K years BP. Since 6K years BP the forest/prairie boundary has shifted westward; its position *c.* 500 years BP was not very different from its location *c.* 9K years BP.

5 The shifts in the prairie/forest boundary are considered to be a response to changing patterns of precipitation and temperatures. These have had both a direct impact on the distribution of vegetation communities, and an indirect impact through their influence on soil moisture deficits (see discussion in Delcourt and Delcourt, 1991).

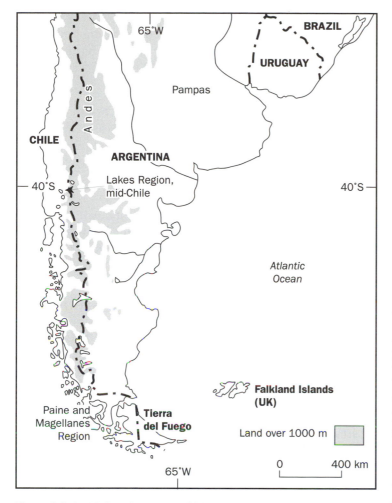

Figure 8.9 *South America, south of 30°S: location of sites and regions referred to in the text*

these occurred. There is evidence for a glaciation at 170K years BP, which was more extensive than the last glaciation (Andersen and Borns, 1994). Clapperton (1994) has, however, produced a detailed reconstruction of the last glaciation (equivalent to the Wisconsin of North America and the Weichselian of Europe) and points out that the glacial deposits indicate at least five advances of glacier lobes from the centres of accumulation. During this period, snow-lines were 800–1000 m lower than at present. In the Chilean lakes region (Figure 8.9) the glaciers reached their maximum extent at *c*. 20K years to 21K years BP; in the Paine and Magellanes region (Patagonia) they were most extensive at *c*. 27K years BP. The latter is attributed by Clapperton to increased humidity which generated snowfall. Ice recession began *c*. 18K years BP, and there are moraines that may have been deposited during the **Younger Dryas** cold period at *c*. 11.5K years BP to 9.5K years BP. Certainly, this cold period is registered in Antarctic ice cores, though there is much controversy as to whether or not the evidence for it, i.e. glacial deposits and pollen sequences in South America, is unequivocal.

Views on the Younger Dryas are polarised, as reflected in the reviews of Heusser (1993) and Markgraf (1993). The former has reviewed pollen-analytical data from six sites in Tierra del Fuego (for location see Figure 8.9). These data contain evidence for two cold periods, one of which occurs between 11K years BP and 10K years BP. In contrast, Markgraf, considering several other pollen diagrams from these southern latitudes, does not countenance cooling during this time;

instead she interprets the pollen changes as a reflection of natural fires, which are evidenced by the presence of charcoal particles. However, the growing body of evidence for a Younger Dryas climatic recession from, for example, parts of the tropical Andes (Section 9.4) and New Zealand (Section 8.7) may reflect its global significance.

Pollen diagrams from various locations in the middle latitudes of South America provide evidence for environmental change during the Holocene. For example, Prieto (1996) has used pollen analysis to examine Holocene change in the pampas, the extensive grasslands that occupy central Argentina (see Figure 8.9 for location). The results of this study are summarised in Table 8.2. The changing character of the early Holocene vegetation is considered to reflect water availability, with conditions similar to those of the present day developing at *c*. 1K years BP. Further south, in Tierra del Fuego, pollen assemblages from peat deposits in the Valley of Andorra are considered by Borromei (1995) to show the early Holocene replacement of dry steppe, with a low tree density, by the spread of southern beech (*Nothofagus*) forests as the climate ameliorated. By *c*. 5K years BP, a closed forest with peatlands had developed as the humidity increased and, possibly, the temperatures decreased.

The middle latitudes of South America have also yielded valuable information on environmental change during the Holocene via tree-ring records. Some of the available data are discussed in Section 6.2 and in Mannion (1997b).

Table 8.2   *Generalised Holocene environmental change in the Pampas of Argentina*

Age (K years BP)	Vegetation type
0–1	Humid, temperate conditions similar to those of the present day developed
1–5	Subhumid conditions returned, as did vegetation communities similar to those of the pre-10.5K year period
5–8	Grassland developed in the central region and persisted until 7K years BP; in the SW region it persisted until 5K years BP
8–10.5	Pond, swamp and floodplain communities developed, reflecting the presence of more abundant moisture. Possibly precipitation concentrations had increased to values similar to those of today
Pre-10.5	Herbaceous **psammophytic** * and **Halophytic**† steppe developed in the central pampas; in the SW region **xerophytic** ‡ woodland occurred in association with the steppe communities. A subhumid climate with an annual precipitation *c*. 100 mm lower than at present predominated.

* Sand loving.
† Salt loving.
‡ Drought tolerant.

Source: Based on Prieto (1996).

8.6 South Africa

The line of latitude 30°S lies just south of the city of Durban and separates Namaqualand, Griqualand and the Orange Free State from the Cape provinces of the rest of South Africa (Figure 8.10 for location). In a recent review of environmental change in this region, Partridge (1997) states that: 'In common with most other midlatitude regions, southern African environments responded dramatically to the global episode of cooling and drying between 2.8 and 2.6 myr [10^6 years] which ushered in the cyclical fluctuations of the Pleistocene'. He also points out that the combination of global climatic change

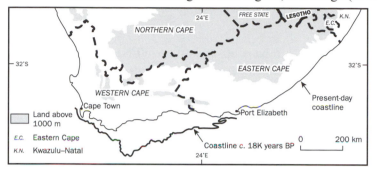

Figure 8.10 *South Africa, south of 30°S and the likely coastline at c. 18K years BP*

Source: Based on Van Andel (1989).

and tectonic uplift in the southernmost region was responsible for the initiation of a winter rainfall regime and the development of the rich scrub flora known as the Fynbos.

However, there is no evidence for direct glaciation in this region and there is little information available from which to determine the impact of episodes of global cooling on South Africa's environment. In relation to the last 30K years, Talma and Vogel (1992) have reconstructed palaeotemperatures from **speleothem** deposits (Section 5.7) in the Cango Caves of Cape Province. For example, during the last glacial maximum at *c.* 18K years BP, temperatures were 6 to 7°C lower than they are today, and there is evidence for marked temperature variations during the last 5K years. The latter are also considered to have occurred in southern South America (Jerardino, 1995); they comprise temperature declines of 1–2°C and sea-level oscillations of −2 to +2 m during the periods 4.5K years BP to 4K years BP, 3K years BP to 2K years BP, and during the last 1K years. These changes are attributed to slight northward shifts of frontal systems, and strong atmospheric circulation.

During the last glacial maximum, sea-levels were much lower than they are at present. Estimates vary as to precisely how much lower but values of −130 m to −170 m are considered likely. In consequence, continental shelf regions were exposed. In South Africa the coastline may have been located as much as 150 m further south than it is today, as illustrated in Figure 8.10. The transition between the last **glacial stage** and the Holocene appears to have been characterised by a gradual climatic amelioration. Scott *et al.* (1995) have reviewed evidence for environmental change during this period, and although a cooling episode equivalent to the Younger Dryas is reflected in sea surface temperatures (as indicated by marine molluscs), there is no pollen-analytical evidence for land-based cooling of sufficient magnitude to influence vegetation communities.

8.7 Australia and New Zealand

Figure 8.11 shows the area of Australia south of latitude 30°S, and the location of New Zealand. Both regions experienced periods of direct glaciation during the

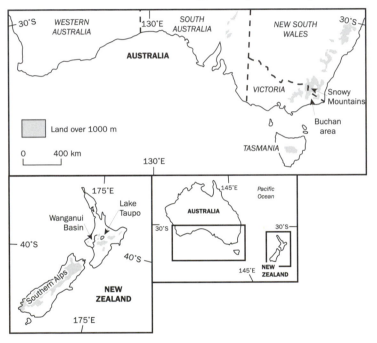

last 3×10^6 years, though it was most extensive in New Zealand because of the presence of the Southern Alps. In Australia, glaciation occurred in the Snowy Mountains of New South Wales (Figure 8.11) and in the Tasmanian Mountains. In addition to glaciations, there is a record of volcanic activity in New Zealand; this has occurred at intervals throughout the last 3×10^6 years. The ash layers so produced provide useful stratigraphic markers as well as material for radiometric dating.

Figure 8.11 *Australasia, south of 30°S: location of sites and regions referred to in the text*

In Tasmania there is evidence for at least four **glacial** episodes. Borehole data from the Boco Plains indicate that glacial advances occurred during **oxygen isotope stages** 6, 8 and 10 (Augustinus *et al.*, 1994). As elsewhere, however, most evidence is available for the last glacial/interglacial cycle, not only from the terrestrial environment but also from ocean sediments. Colhoun *et al.* (1994) have compared glacial records and terrestrial and marine pollen data to reconstruct environmental change during the last *c.* 130K years. The results are summarised in Table 8.3. This also shows the environmental changes that occurred during the same time interval in western Victoria. In both locations, forest developed during the **interglacial stages** (oxygen isotope stages 1 and 5, especially 5e) and during the last ice age there is evidence for **interstadial** conditions that were climatically milder and wetter than the rest of the period.

Table 8.3 *A summary of environmental change in Tasmania (based on Colhoun et al., 1994 and Augustinus et al., 1994) and western Victoria (based on Harle, 1997) during the late Quarternary*

Oxygen isotope stage	Environmental characteristics TASMANIA
1	Holocene warming; forest development
1–2	No evidence of younger Dryas cooling
2	Last glacial maximum occurred between 25K and 19K years BP; environment similar to/harsher than stage 4
3	Sub-Alpine woodland/shrubland; a milder climate than stages 2 or 4
4	Ice advance; herb/shrub vegetation beyond ice limits, reflecting a cold, dry climate
5	Wet forest; warm, wet climate
6, 8, 10	Glacial advances

Oxygen isotope stage	Environmental characteristics WESTERN VICTORIA
1	Holocene amelioration; scrub, eucalypt and *Casuarina* woodland developed
2	Decline in effective precipitation; open heath and grass communities developed as a result
3	Interstadial; increased effective precipitation, with limited expansion of *Eucalyptus*
4	Open heath and mallee communities reflect drier conditions
5a/5b	Gradual development of drier conditions and replacement of wet forest
5c	Return of wet forest; high effective precipitation
5d	Dry sclerophyll forest; decrease in effective precipitation
5e	Wet sclerophyll forest and rain forest; high effective precipitation

In relation to the last glacial/cold stage transition to the Holocene, there is little evidence for a climatic excursion equivalent to the **Younger Dryas**. However, recent dating of growth stages for a calcitic stalagmite from a cave in the Buchan area of eastern Victoria has shown that notably cool conditions occurred between 12.3K years BP and 11.4K years BP, and again at *c.* 3K years BP (Goede *et al.* 1996). The earlier episode is synchronous with the European Younger Dryas and corresponds with a similar climatic regression recorded in the Vostok (Antarctic) ice core (Section 4.5).

In recent years, the record of environmental change in New Zealand has received considerable attention. Marine-sediment records, glacial and periglacial deposits in the terrestrial environment and volcanic deposits all attest to the dynamism of environmental change in the last 3×10^6 years. Some of the most extensive deposits representing this period occur in the Wanganui Basin (Figure 8.11), in the southern part of North Island. They comprise marine terraces, **loess**, dune sand and bands of rhyolite tephra. According to Shane *et al.* (1996) there are 54 different tephra layers present in the basin. These reflect the occurrence of volcanic eruptions in the Taupo volcanic zone (the Lake Taupo area, see Figure 8.11 for location) between 2×10^6 years and 600K years BP, with particularly intense volcanism between 1.79×10^6 and 1.6×10^6 years ago. The ages of these tephra horizons have been estimated radiometrically, and thus provide a means of establishing a chronostratigraphy, and of correlation between terrestrial and marine deposits. Pillans (1991) has described the sedimentary sequence deposited during the last 2.6×10^6 years and has related the various deposits that accrued during the last 1×10^6 years in order to reconstruct environmental change (Pillans, 1994). The latter is illustrated in Figure 8.12. This shows that during warm stages marine terraces were formed along with coastal sand dunes. During cold stages, loess was deposited.

In relation to the Younger Dryas event, there is some evidence from the South Island for glacial re-advance at $c.$ 11K years BP (Fitzsimons, 1997), but it remains controversial. There is also evidence for glacier re-advance during the Little Ice Age.

8.8 Conclusions

In some mid-latitude regions there are key sites for the elucidation of environmental change during the last 3×10^6 years, notably the deep lacustrine sequences of Japan, Israel and the **loess** sequences of Asia. Such deposits not only provide a record of local and regional environmental change, but also facilitate correlations with the oxygen isotope stratigraphy of marine-sediment cores. Consequently, they make an important contribution to the establishment of trends of global environmental change. Overall, such records reflect the operation of Milankovitch cycles of climatic change (Section 2.4; Box 2.2).

These lacustrine and loess sequences occur in the middle-latitude regions that did not experience direct glaciation. In the Northern Hemisphere in particular many areas were ice covered, e.g. Canada and the Northern USA, at many different times in the past 3×10^6 years. Evidence for this and intervening warm periods includes glacial deposits and palaeosols. Together with ocean-sediment records, this evidence confirms the operation of numerous glacial–interglacial cycles during the last 3×10^6 years. Inevitably, the most detailed information is available for the most recent climatic cycle, and evidence from a range of archives reflects the rapid and complex shift from glacial to interglacial conditions.

Evidence relating to earlier **interglacial stages** indicates that the last interglacial (**oxygen isotope stage** 5e) was generally warmer by $c.$ 1°C–2°C than the present interglacial. This is reflected in the fossil assemblages of foraminifera (Figure 3.3) in ocean sediments, and in the pollen assemblages of terrestrial deposits.

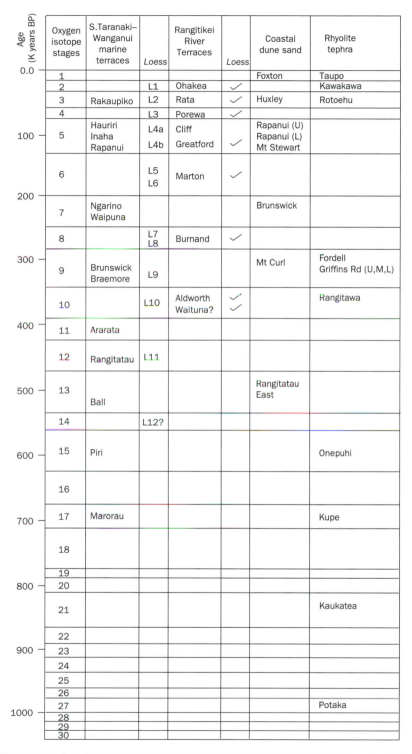

Age (K years BP)	Oxygen isotope stages	S.Taranaki–Wanganui marine terraces	Loess	Rangitikei River Terraces	Loess	Coastal dune sand	Rhyolite tephra
0.0	1					Foxton	Taupo
	2		L1	Ohakea	✓		Kawakawa
	3	Rakaupiko	L2	Rata	✓	Huxley	Rotoehu
	4		L3	Porewa	✓		
100	5	Hauriri Inaha Rapanui	L4a	Cliff		Rapanui (U) Rapanui (L) Mt Stewart	
			L4b	Greatford	✓		
	6		L5 L6	Marton	✓		
200	7	Ngarino Waipuna				Brunswick	
	8		L7 L8	Burnand	✓		
300	9	Brunswick Braemore	L9			Mt Curl	Fordell Griffins Rd (U,M,L)
	10		L10	Aldworth Waituna?	✓ ✓		Rangitawa
400	11	Ararata					
	12	Rangitatau	L11				
500	13	Ball				Rangitatau East	
	14		L12?				
600	15	Piri					Onepuhi
	16						
700	17	Marorau					Kupe
	18						
	19						
800	20						
	21						Kaukatea
	22						
900	23						
	24						
	25						
	26						
1000	27						Potaka
	28						
	29						
	30						

Figure 8.12 *The record of environmental change in Wanganui Basin, North Island, New Zealand*
Source: Based on Pillans (1994).

The middle latitudes of the Southern Hemisphere have been the subject of much recent investigation, even though the land area is much less extensive than in the Northern Hemisphere. Ocean sediments, and terrestrial deposits from South Africa, Australia and New Zealand confirm results from elsewhere, notably that the climatic cycles were just as pronounced in the Southern Hemisphere as in the Northern Hemisphere.

Summary Points

- The middle latitudes provide a wealth of evidence for environmental change over the last 3×10^6 years.

- The most extensive evidence is in the Northern Hemisphere because of the greater extent of the land area.

- In formerly glaciated middle latitudes, glacial deposits and palaeosols attest to past environmental change.

- In extraglacial middle latitudes, **loess** and lacustrine sequences are key elements in the reconstruction of environmental change.

- The sediments of Lake Biwa, Japan, extend back $c.$ 6×10^6 years into the Tertiary period.

- Long lake sequences from Greece, Italy, Israel and Lake Baikal provide evidence for environmental change over the last 1×10^6 years.

- Correlations are possible between terrestrial deposits and **marine oxygen isotope stages**.

- There is a growing volume of evidence for environmental change from the Southern Hemisphere, which experienced the same intensity of climatic change as the Northern Hemisphere.

General further reading

Environmental Change. A.S. Goudie. 1992. Clarendon Press, Oxford, 3rd edn.

Ice Age Earth. A.G. Dawson. 1992. Routledge, London.

The Ice Age World. B.G. Andersen and H.W. Borns Jr. 1994. Scandinavian University Press, Oslo, Copenhagen and Stockholm.

The Pleistocene. Geology and Life in the Quaternary Ice Age. T. Nilsson. 1983. Reidel, Dordrecht.

Quaternary Environments. M.A.J. Williams, D.L. Dunkerley, P. De Deckker, A.P. Kershaw and T. Stokes. 1998. Arnold, London, 2nd edn.

9 Environmental change in low latitudes (30°N and 30°S)

9.1 Introduction

Traditionally, most research on environmental change has focused on middle- and high-latitude regions. This is because such regions were directly affected by glacial and/or periglacial processes, as discussed in Chapters 7 and 8. Moreover, the absence of glaciation from all but the highest regions of low latitudes has led to the belief that such regions had remained largely unaltered throughout the last 3×10^6 years. The advent of palaeoenvironmental information from ocean sediments, beginning in the 1950s, scotched this myth as theories of global-scale change began to be formulated. The recognition that global temperatures were depressed by $c.$ 10°C during the ice ages, prompted questions about the nature and direction of environmental change in extraglacial regions, especially within the tropics.

Have the world's deserts always enjoyed their present distribution, and have their boundaries shifted as climatic change has occurred? How have the world's most productive vegetation communities, notably tropical forests, been affected by temperature depression and elevation? These are just some of the questions that palaeoenvironmental research in low latitudes has addressed.

The evidence for environmental change during the last 3×10^6 years derives from a range of archives, as it does in middle and high latitudes. Evidence from ocean cores provides the longest and most complete record of environmental change (Chapter 4) in low latitudes. However, there are a number of long lacustrine sequences that provide evidence of changing conditions on the continents. The longest of these is from Funza, near Bogotá in Colombia, which extends back to the Pliocene period (Section 5.3.1). Glaciers, such as Huascarán in the Peruvian Andes, also provide palaeoenvironmental information, though peat and mire deposits are the most common sources of data on environmental change. In particular, pollen analysis of such deposits is making a major contribution to the elucidation of environmental change.

9.2 Africa

Parts of the highland zone of Africa were sufficiently cooled at regular intervals during the last 3×10^6 years to experience direct glaciation. In the case of Mount Kenya, for example, the glacial history has been determined, and is discussed in Box 9.1. The advance and retreat of Mt Kenya's glaciers was paralleled by shifts in Africa's vegetation belts, evidence for which includes pollen analysis of ocean-sediment cores of the north-west coast of Africa. A 200 m core from ODP (Ocean Drilling Programme) Site 658 (see Figure 9.1 for location) provides a record encompassing the period 3.7×10^6 to 1.7×10^6 years. The pollen assemblages show that by $c.$ $3.26 \times \times 10^6$ years BP, *Ephedra* pollen had increased substantially; this

---

Box 9.1

A summary of the environmental history of Mt Kenya during the last 1×10^6 years

Deposits (e.g. tills)	Approximate age K years BP	Oxygen isotope stages
Liki	begins c. 100	4
Teleki	c. 200	6
Naro Moru	pre 320	
Lake Ellis	pre 730 to 790	
Gorges		
Pre-Gorges		

1 As is the case in high and middle latitudes, Mt Kenya, along with other tropical mountains, experienced multiple glaciation.

2 There is limited evidence for early Quaternary glaciation, but during the middle and late Quaternary period there were five or six major ice advances. The tills listed above have palaeosol development on their surfaces; many are highly weathered and thus probably formed during interglacial stages.

3 There is evidence that the last ice age (Liki) was characterised by three major ice advances, dated at $c.$ 80K, 20K and 11K years BP. The latter reflects the Younger Dryas event. There is also evidence for neoglaciation during the last 1K years BP.

Source: Based on Mahaney (1990)

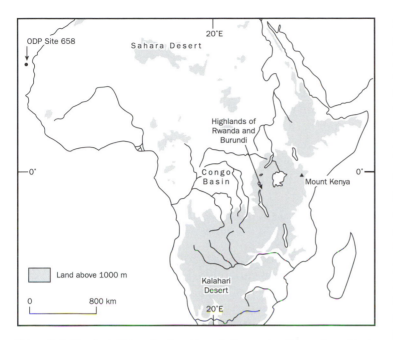

Figure 9.1 *Tropical Africa: the location of sites and regions referred to in the text*

reflects the development of arid conditions, and possibly the establishment of the trade winds (Leroy and Dupont, 1994). It also coincides with the onset of glaciation in the North Atlantic (Sections 2.3; 7.2). There is also the possibility that a further shift towards increasingly arid conditions contributed to **hominid evolution** (Demenocal, 1995). In addition, the pollen assemblages of ODP Site 658 reflect the alternation of humid conditions with arid conditions, which caused latitudinal shifts of wooded savannah and desert. The latter correspond with even-numbered **oxygen isotope stages**, i.e. cold stages, whilst savannah vegetation communities developed during the warm stages. Dupont (1993), citing the results of pollen analysis of a further six cores from the East Atlantic, suggests that during cold stages the Sahara Desert expanded. For example, the southern boundary may have been located at *c*. 14–15°N, in contrast to its position at *c*. 20°N today (Figure 9.1). Dupont also points out the sensitivity of this boundary (the Sahara–Sahel) to climatic change in the past, a sensitivity that may well be reactivated, with considerable consequences for human life, by anthropogenic global warming. In contrast, the Kalahari Desert (for location see Figure 9.1), in the Southern Hemisphere, experienced increased precipitation during the 16K and 13K year BP period, after which arid conditions characteristic of the present day developed (Shaw and Thomas, 1996). Further research on environmental change in the Kalahari has also shown that the last cold stage experienced climatic variability akin to the occurrence of interstadials in high and middle latitudes. For example, arid dune-building phases have been identified from the Mega Kalahari sand sea; these occurred during the intervals 95K to 115K years BP, 41K to 46K years BP, 20K to 26K years BP and 9K to 16K years BP (Stokes *et al.*, 1997).

Increased aridity during the last cold stage in tropical Africa is also reflected in a series of pollen diagrams from the highlands of Burundi, Rwanda and western Uganda (Jolly *et al.*, 1997). Here (see Figure 9.1 for location) *c*. 18K years BP ericaceous scrub (heathland species similar to heather), and grasslands with patches of open montane forest comprised the dominant vegetation communities. By *c.* 12.5K years BP, montane forest characteristic of upper altitudes was expanding markedly, whilst forest characteristic of the lower montane region did not expand fully until 11K to 10K years BP. According to this and other work in the region, the colonisation by forest was not uniform either spatially or temporally. Moreover, the pollen record and carbon accumulation rate of a peat deposit at Rusaka in Burundi both indicate that a climatic regression occurred between 10.6K years and 10K years BP. Bonnefille *et al.* (1995) suggest that this is equivalent to the **Younger Dryas** cold period, which is recorded in a wide range of archives at various latitudes and altitudes (e.g. Sections 4.5, 8.2 and 9.4). At Rusaka, however, the interruption of forest development and opening of the canopy is considered to be the result of increased aridity rather than a temperature decline.

Palaeoenvironmental evidence from numerous sites in tropical Africa indicates that the Holocene was characterised by variability. For example, from 5.3K years BP in the eastern Sahara, Neumann (1991) has demonstrated that aridity increased. This resulted in a shift southward of savannah communities by *c.* 500–600 km as the desert margins expanded. This ceased at *c.* 3.3K years BP, by which time the distribution of savannah and desert characteristic of the present day had developed. There is also pollen-analytical evidence, notably a decline in tree pollen and an increase in grass pollen, for increased aridity *c.* 3K years BP in the Congo (Elenga *et al.*, 1994). The decline in precipitation caused a decline in forest extent and possibly fragmentation, as is characteristic of modern plant communities in the region. Increased aridity is also considered to have occurred between 4K years and 1.2K years BP around Lake Sinnda in the Guineo-Congolian region (Alexandre *et al.*, 1997). Changing patterns of precipitation have clearly influenced tropical African environments during the late Holocene.

9.3 Asia

The low latitudes of Asia between 30°N and 30°S (Figure 9.2) include some of the highest mountains in the world, notably the Himalayas. In common with the European Alps (Box 8.1) and highland Africa (Box 9.1), the Himalayas experienced direct glaciation during the last 3×10^6 years. Since this time, mountain uplift has continued, and may have

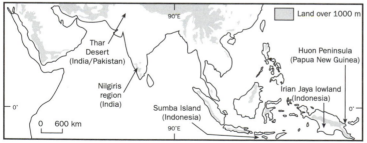

Figure 9.2 *Tropical Asia: the location of sites and regions referred to in the text*

contributed to the onset of ice age/interglacial climatic cycles, as discussed in Section 2.4. There is considerable debate as to how many **glacial stages** occurred in the Himalayas and their nature, i.e. extensive ice-cap development or dominance of glaciers emanating from localised montane ice caps. The possibilities are considered in Box 9.2.

Further south, in India, sedimentary sequences in the Thar Desert of Rajasthan provide a means of reconstructing Quaternary environments. According to Sundaram *et al.* (1996) variations in the sediments of a number of basins, including pedogenic and geomorphological characteristics, reflect oscillations between humid tropical and arid conditions. For example, at the onset of the Quaternary period, arid conditions prevailed; thereafter humid and arid conditions alternated in tandem with ice ages and **interglacial stages**. Similar oscillations occurred in the Sahara Desert (Section 9.2). Phases of aridity and humidity have also been detected in tropical peats of the Nilgiris montane region of southern India. The peats began to form sometime during the last cold stage; the earliest age estimation is *c.* 40K years BP (Rajagopalan *et al.*, 1997). The record of stable carbon isotopes reflects the proportions of various plant types, which in turn are influenced by soil moisture. Using such data, the sequence of events given in Table 9.1 is considered to have occurred during the last 40K years.

Table 9.1   *Environmental change in the Nilgiris montane region, southern India, during the last c. 40K years*

Age estimate (K years BP)	Conditions
<40–28	Increasing moisture
28–18	Little change: moist conditions prevailed
c. 16	Rapid development of arid conditions during or just after the last glacial maximum
9–5	Return to moist conditions
5–2	Arid conditions re-established

Source: Based on Rajagopalan *et al.* (1997).

Changes in sea-level paralleled the accumulation and demise of major ice sheets. In particular, Wang and Sun (1994) have discussed the impact of the drop in sea-level at the last glacial maximum *c.* 18K years BP on the South and East China Seas. In the former, sea-level fell by 100–120 m, and in the latter the decline was *c.* 130–150 m. As shown in Figure 9.3 the configuration of the coastlines changed substantially. In the South China Sea, which became a partially enclosed gulf, winter temperatures were between 6 and 10°C colder than at present, with an increased seasonal range. This was caused by the movement south of the polar front and the alteration of surface currents due to lowered sea-level. Wang and Sun also provide evidence for increased aridity in continental China. As a result, **loess** deposition became extensive, including in the region south of the Yangtze River (Figure 9.3). Furthermore, vegetation belts shifted to the south-east. In south China, for example, the monsoon rain forest was absent and deciduous broad-leaved forests developed in areas now occupied by evergreen forests; both these shifts were a response to increasing aridity.

Box 9.2

A synopsis of the environmental history of the Qinghai–Xizang (Tibet)–Himalaya region

1 The Qinghai–Xizang (Tibet) Plateau and the Himalayas are considered to be influential in Tertiary/Quaternary climatic change. Uplift of this region may have contributed to the establishment of climatic cycles comprising cold stage/ice age and warm stage/interglacial conditions (see Section 2.4).

2 The altitude of the region (Mt Everest rises to 8818 m above sea level, the plateau is *c*. 4000 m above sea level and Lhasa, the capital of Tibet, is at 3683 m above sea level), means that it was a likely centre of ice accumulation during the cold stages of each climatic cycle.

3 There is, however, considerable debate about the number, nature and timings of glaciations in this and other parts of the Himalayas (see e.g. Derbyshire, 1996).

4 In general, four glaciations have been recognised in the Qinghai–Xizang (Tibet) Plateau, and the likely chronology, etc., is given below. There is little evidence to support the existence of extensive ice sheets at any time during the last 2.4×10^6 years.

Age (K years BP)	Stage	Evidence
	Holocene	
10		
	Baiyu Glaciation (*Qomolanga II*)	End moraines; evidence for interstadials
70		
	Last interglacial	Various deposits
130		
	Guxiang Glaciation (*Qomolanga I*)	End and lateral moraines, Jilongi
	Great Interglacial	Red palaeosol
	Nyanyaxungla Glaciation	Moraine terrace 5200–5700 m above sea level Nyanyxungla till platform
730		
	First interglacial	Aga palaeosol
	Xixabangma Glaciation	High moraine platform Xixabangma Mountain
2400		
	Pliocene	

Source: Based on Zheng and Rutter (1998)

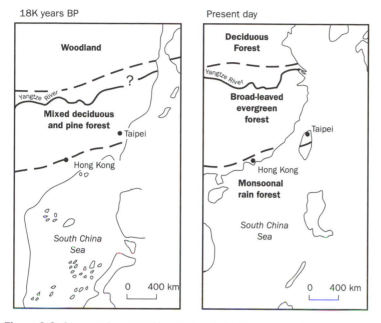

18K years BP

Woodland

?

Yangtze River

Mixed deciduous
and pine forest

● Taipei

Hong Kong

South China
Sea

0   400 km

Present day

Deciduous
Forest

Yangtze River

Broad-leaved
evergreen
forest

●Taipei

Hong Kong

Monsoonal
rain forest

South China
Sea

0   400 km

Figure 9.3 *Changes in the configuration of the South China Sea and the vegetation zones of southern China: the situation c. 18K years BP, as compared with the present*

Source: Based on Wang and Sun (1994).

Sea-level change has also been well documented in parts of the tropics where there are coral reefs. The determination of sea-level change during the last 3×10^6 years is particularly difficult because measurements in one region of the Earth's surface cannot easily be extrapolated to give a measure of global sea-level change. As discussed in Section 1.2.2, sea-level change due to isostatic and eustatic factors has occurred as ice advances have taken place and then regressed. Separating the impacts of these two processes is itself difficult, but to this must be added the complication of **geoidal** or **geodetic eustasy**. This is the influence of the shape of the Earth (a geoid), with its relatively flattened polar regions and bulging equatorial regions. Sea-level changes are, consequently, non-uniform globally. Nevertheless, coral reefs are particularly sensitive indicators of sea-level change, because the coral ceases to grow when the reef is emergent, i.e. when it is at or above sea-level; this may occur during ice ages when the global hydrological cycle is different to that of the present, and a vastly increased volume of water is incarcerated in ice sheets, etc. The records of sea-level change in suites of coral reefs is, however, complicated by several factors, including **isostatic** recovery of the Earth's crust following deglaciation and the release of the overriding weight of ice, and **tectonic** movement related to plate tectonics (Box 1.2). However, the records, including age determination, of two flights of coral terraces in New Guinea and Indonesia provide important information on sea-level change during the later part of the Quaternary period, as illustrated in Figure 9.4A. This shows that the highest sea-level during the last 140K years occurred during **oxygen isotope stage** 5e (the last **interglacial**) when it was *c.* 6 m higher than at present. Such a finding confirms evidence from numerous palaeoenvironmental archives that the last interglacial was warmer, by 1–2°C, than the optimum of the present interglacial (the Holocene). A similar but longer sequence, which covers the last 400K years, has been examined from Sumba Island, Indonesia (Bard *et al.*, 1996). This record is given in Figure 9.4(B). The Sumba record indicates that during the

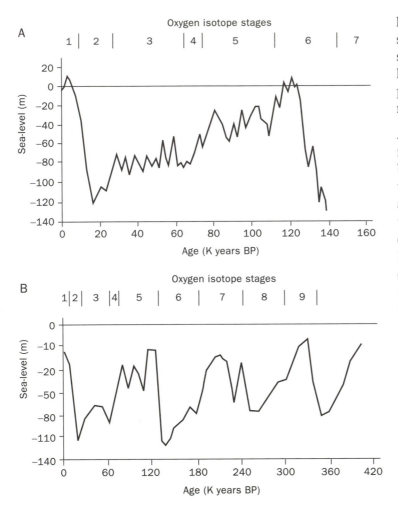

Figure 9.4A *Sea-level changes during the last c. 140K years, on the basis of coral terrace sequences in the Huon Peninsula, New Guinea*

Source: Based on Chappell et al., (1995); oxygen isotope stages derived from Pacific deep-sea core V19-30 reported by Shackleton et al. (1983).

Figure 9.4B *Sea-level changes during the last c. 390K years, on the basis of the coral terrace sequences of Sumba Island, Indonesia*

Source: Based on Bard et al. (1996).

last three interglacial stages, i.e. oxygen isotope stages 5e, 7 and 9, the sea-level was at least as high as present, and possibly as much as 10–20 m higher.

Although the tropical regions of Asia have not been as well investigated as their counterparts in high and middle latitudes in terms of their environmental history, it is likely that long records will emerge from lake-sediment and other archives in the next few years. These will complement existing long records from the ocean sediments (Chapter 3, e.g. Figure 3.1). An example of such a record is that from a mire in lowland Irian-Jaya, Indonesia. Pollen analysis of a 10 m core (Hope and Tulip, 1994) has revealed that for *c.* 60K years, forest has occupied the region but has changed in character. For example, the taxa characteristic of higher altitudinal locations (when compared with present-day plant distributions) increased between 25K and 10.5K years. This reflects a temperature depression of 3 to 4°C, and is consistent with the presence of extensive ice sheets in high and middle–high latitudes. At *c.* 10.5K years BP, taxa characteristic of lower altitude forest began to invade, with a possible reversal at *c.* 10.2K years BP. The latter may reflect changing conditions at the time of the Younger Dryas event in middle and high latitudes. By *c.* 7K years BP, low-altitude forest had become well established. This record reflects the continuity of forest cover over the last 60K years, but also provides evidence for changes in composition as the taxa reacted to climatic change.

9.4 The Americas

In recent years, a considerable volume of research data has emanated from the low latitudes of the Americas. This has involved the investigation of many archives of palaeoenvironmental data, but especially lake sediments and some peats. Thus there has been a concentration on indicators specific to those environments, notably pollen analysis (Box 5.1). In particular, this technique has contributed to the understanding of the vegetation dynamics (Section 1.2.3 in relation to theories of succession and climax) of tropical forests.

One of the most remarkable records of environmental change is that of Funza, from the High Plain of Bogotá, Colombia (see Figure 9.5 for location). This has been referred to in Section 5.3.1, and Figure 5.6 gives a record of changes in the percentage arboreal pollen for the last c. 850K years. However, the record extends back to 3.5×10^{6} years BP, and involves some 27 glacial–interglacial cycles (Hooghiemstra, 1989), as shown in Figure 9.6. If the altitudinal limit of the forest is a crude proxy measure of the prevailing temperatures, the longest cold period of the last 3.5×10 years occurred c. 1.7×10^{6} years (during **oxygen isotope stage** 24) at the opening of the Quaternary period, and the warmest periods, when the forest was at its highest altitude, correspond with oxygen isotope stages 7 and

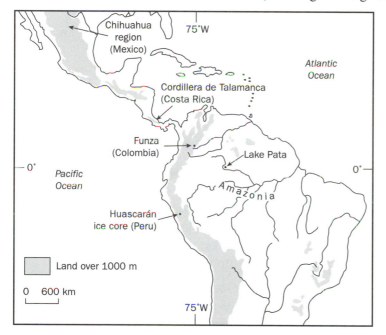

Figure 9.5 *Tropical America: the location of sites and regions referred to in the text*

11, i.e. 250–200K years BP and 440–360K years BP. This interpretation may, however, be too simplistic, as factors such as water availability and evaporation rates are also important determinants of tree-line location. Nevertheless, the pollen record is unequivocal in reflecting the dynamism of vegetation in the region over the last 3.5×10^{6} years.

There are no other records from tropical America that are equivalent temporally to the Funza sequence. Moreover, the paucity of palaeoenvironmental records from areas such as Amazonia has created a situation where a number of hypotheses have been presented, but these are only now beginning to be tested

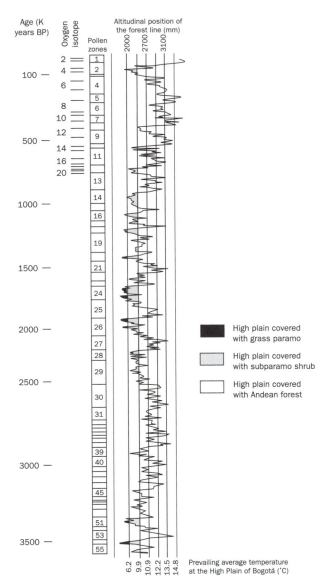

Figure 9.6 *The relationship between climatic change, as represented by marine oxygen isotope stages, and changes in the altitudinal position of the tree-line in the High Plain of Bogotá in the Eastern Cordillera of Colombia, as deduced from the Funza I. core*

Source: adapted from Hooghiemstra (1989).

adequately. For example, ecologists have suggested that during the ice ages, Amazon rain forest was greatly reduced in area, to a number of isolated fragments, i.e. **refugia**. These were surrounded by vegetation communities of a very different type, notably savannahs and/or grasslands. Within these refugia, speciation occurred to increase biodiversity, and these old and new species subsequently colonised the former savannahs and/or grasslands when the climate ameliorated during the **interglacial stages**. This so-called **refuge** hypothesis (one of its original proponents was Haffer, 1969) is thus often advanced as an explanation for the high biodiversity of tropical rain-forest communities (Section 8.2). Consequently, it has generated considerable debate. Pollen analysis is a suitable tool for testing this hypothesis, providing that suitable archives of pollen data can be found. Where pollen analysis has been undertaken the record does not support the refuge hypothesis.

One such site is Lake Pata (see Figure 9.5 for location), which is today surrounded by dense tropical rain forest. The catchment of the lake is not considered to be a refugium; consequently, the pollen record would be expected to show a predominance of savannah/grassland types in sediments deposited before about 15K years BP, i.e. during ice age times when increased aridity is considered to have occurred, and then a return to tree taxa during the Holocene. However, Colinvaux *et al.* (1996) have shown that throughout the 7 m core, the pollen of tropical rain-forest taxa predominated from *c.* 42K years BP to present; savannah and/or grassland never occupied Lake Pata's catchment. Although this is one of few sites investigated in Amazonia, and according to Colinvaux *et al.* it is the first to extend back into the period of the last glacial, the results do not support the possibility of tropical rain-forest fragmentation. By inference, this undermines the refuge hypothesis. The pollen spectra do, however, show changes over time. Indeed, the pollen data from Lake Pata are in some ways similar to those from Irian Jaya in Indonesia (Hope and Tulip, 1994; Section 9.3). At both sites, pollen of taxa characteristic of present-day higher altitudes increased in the sediments deposited during the last ice age. Colinvaux *et al.* (1996) believe that this pollen record of Lake Pata indicates a temperature depression of *c.* 5–6°C, whilst Hope and Tulip (1994) suggest a temperature depression of 3–4°C for Irian Jaya. Thus, although temperatures declined, there is no evidence in the pollen record for increased aridity or forest fragmentation. Studies such as these are important not only for their contribution to an understanding of the dynamics of tropical vegetation communities, but also for their contribution to tropical palaeoclimatology. This – albeit limited – evidence for temperature change during the last ice age in the tropics, implies that the depression was not as great in the lowland tropics as it was in the Andes, as evidenced by the Funza core, which is interpreted as indicating a *c.* 9°C temperature difference between glacial and interglacial stages, and in middle and high latitudes (Chapters 7 and 8).

Another ecological issue to which tropical palaeoecology has contributed is the question of whether vegetation communities reacted to climatic change as whole communities or in terms of the component species. This issue relates to the debate between plant associations, in the sense defined by Tansley, and the individualistic concept of the plant association, as advocated by Gleason in the early twentieth century (Section 1.2.3). The debate is epitomised by the question: does climatic change cause entire communities to react *in toto* spatially, or do individual taxa alone group and regroup to create unique communities spatially and temporally? In other words: do plant communities remain constant in time and space and react as entities to climatic change, or do individual taxa react to such stimuli to create novel plant communities? Most opinion now leans towards the latter view (Section 5.8).

Ice-core research is also providing important information about environmental change, particularly climatic change. Most recently, Thompson *et al.* (1995) have extracted cores from Huascarán in Andean Peru (see Figure 9.5 for location). A summary of the results is given in Box 9.3. Apart from the fact that the data provide evidence for local environmental change, many trends apparent in the Huascarán cores parallel trends in polar ice cores (Chapter 4) and archive data from other parts of the world. For example, there is clear evidence for a **Younger Dryas** cold stage in the Huascarán oxygen isotope stratigraphy (Box 9.3). There is much debate as to whether or not this is a global event, and whether or not it is represented in tropical America (reviewed in Mannion, 1997b). The Huascarán data provide unequivocal evidence for a temperature decline during this period, but elsewhere in the region evidence for Younger Dryas cooling is limited. Islebe

Box 9.3

A summary of the palaeoenvironmental data and inferred conditions from the Huascarán ice cores, Andean Peru

1 The Huascarán ice cores were obtained from glaciers in the north-central Andes of Peru in the Cordillera Blanca at latitude c. 9°S (for location see Figure 9.4). The cores were 160.4 m and 166.1 m in length.

2 A variety of parameters were measured, including oxygen isotope ratios (see Box 3.1) and deuterium concentrations (see Section 4.3.2 for an explanation of principles). Such records can be compared with those from polar ice cores (Chapter 4).

3 The Huascarán ice cores provide a record for the last c. 20 K years: the end of the last ice age, the glacial–interglacial (Holocene) transition and the Holocene.

4 There is oxygen isotope evidence for the Younger Dryas cool interval and changing environmental conditions associated with it. These are summarised below.

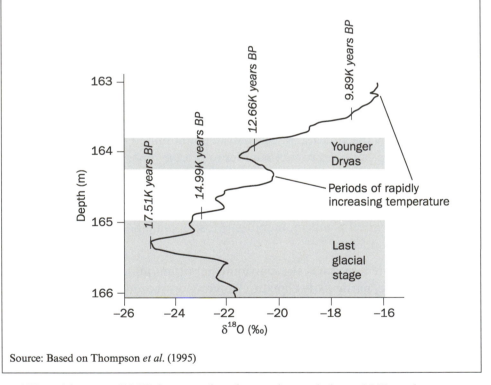

Source: Based on Thompson *et al.* (1995)

and Hooghiemstra (1997) have analysed cores located above 2300 m above sea-level, from the Cordillera de Talamanca in Costa Rica. They interpret the changes in the pollen assemblages as a reflection of a decline of 7–8°C during the last glacial maximum (c. 18K years BP) and a decline of 1.5–2.5°C during the Younger Dryas (c. 12K to 10K years BP). They also cite similar evidence from

Guatemala. In addition, increased aridity is considered to have occurred during the Younger Dryas in closed lake basins 1280–2200 m above sea-level in the Chihuahua region of Mexico (Metcalfe *et al.*, 1997). Although, overall, relatively few sites have been investigated in tropical America, there appears to be positive evidence for the Younger Dryas climatic regression in upland regions but limited supporting data from lowland regions. Possibly, the latter did not experience a temperature depression of sufficient magnitude to initiate significant vegetation change.

9.5 Australia

Some of the longest records of environmental change in Australia derive from the humid tropical region of north-eastern Queensland. Here, in the Atherton Tableland (for location see Figure 9.7) there are a number of former volcanic craters in which sediment has accumulated. For example, Lynch's Crater contains a record of the last *c.* 190K years (Kershaw, 1986). This is just *c.* 30K years more than the record in the Vostok (Antarctica) ice core (Chapter 4, especially Figure 4.2). Pollen analysis of the sediments provides a record of vegetation change; these data, together with pollen data from Strenekoff's Crater and a marine sedimentary sequence (ODP Site 820) off the coast of north-east Queensland, have been

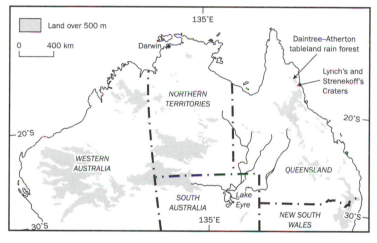

Figure 9.7 *Tropical Australia: location of sites and regions referred to in the text*

used by Kershaw (1994) to produce a palaeoenvironmental record for the last *c.* 2×10^6 years. The pollen data show that, in general, **glacial/cold stages** were characterised by relatively dry tropical vegetation communities, notably **sclerophyll woodland** (including *Eucalyptus* spp.) and Araucarian rain forest. The latter is so called because of the dominance of the genus *Araucaria* (various pines) along with another conifer, *Podocarpus* (some species of which are also referred to locally as pine). This is in contrast to the **interglacial/warm stages**, when a rain forest dominated by **angiosperms** prevailed. However, Kershaw (1994) concludes that the most significant change in vegetation has occurred in the last 140K years. Moist rain forest has been replaced by open eucalyptus woodland, but because this has not been synchronous between sites it is possible that it has been brought about by human activity. There is clear evidence of burning, as reflected in the

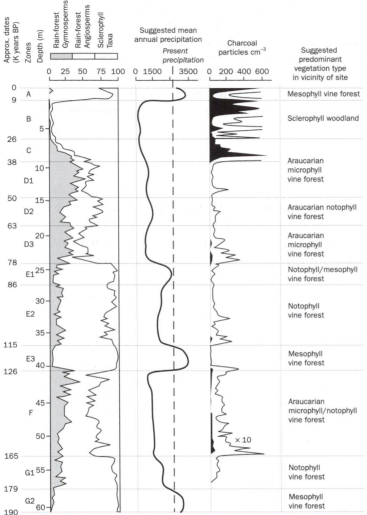

Figure 9.8 A summary of the results of pollen analyses from Lynch's Crater, Queensland, Australia

Source: Based on Kershaw (1994).

charcoal record of Lynch's Crater (Figure 9.8) and although this may have been produced by natural fire, it could equally well have been produced by the burning of vegetation by Aboriginal people.

There is growing evidence that the environmental changes, prompted by climatic change during the last 3×10^6 years, have influenced the genetic characteristics of faunal populations in Australia's tropical forests. For example, Joseph et al. (1995) have examined the genetic characteristics of several bird and one lizard species living in two tropical forest areas in north-east Australia. The rain forests of the Atherton Tableland and Daintree (for location see Figure 9.7) are currently contiguous. This was not the case during the last ice age/cold stage, when the area of rain forest shrank and divided into two distinct units, separated by the Black Mountains. The genetic analysis of these species included the analysis of a genetic component which accumulates mutations at a regular rate and can thus be used as a molecular clock. Four of the five bird species exhibited substantial differences in this characteristic and can thus be considered as reflecting the development of two distinct populations, which have since been united as the forest has coalesced. In the case of the lizard (the prickly skink), the genetic data show differences that may be interpreted as the beginning of divergence into two distinct species. These data imply that the geographical isolation of populations may lead to species divergence. Thus the glacial stages of the past 3×10^6 years may well have stimulated evolutionary processes.

There is also evidence for environmental change in Australia's arid zone e.g. the Lake Eyre Basin. Magee *et al.* (1995) report that during the last interglacial stage the lake was at its deepest: *c.* 25 m. Thereafter, aridity characterised the last glacial stage. In addition, many of the fossil dune-fields of the interior began to form *c.* 20-24K years BP, i.e. prior to the last glacial maximum. This was also a time of increasing aridity in Australia as lake levels declined (Harrison and Dodson, 1993). Such findings are in accord with the dust record, i.e. the aluminium record (Figure 4.6) in the Vostok ice core; the concentration of aluminium, considered to be wind blown (it was probably attached to eroded soil/silt particles), increased substantially during the last ice age.

The configuration of Australia's coastline was, inevitably, quite different during the last ice age at *c.* 18K years BP. The decline in sea-level that accompanied the expansion of the polar ice sheets caused sea-level to decline by between 135 m and 175 m. This, in turn, increased the land area of Australia by *c.* 20 per cent, and exposed land in the seas to the north, so that a land bridge between northern Australia and Papua New Guinea came into existence, as shown in Figure 9.9. This is discussed by Williams *et al.* (1993), who suggest that this land bridge altered the climate in Australia by diverting the warm south equatorial current, which flows between the two islands at present. Consequently, the aridity was enhanced in northern Australia as it no longer received moist air from the ocean. This increase in aridity is reflected in the vegetation changes of the Atherton Tableland, as discussed at the beginning of this section.

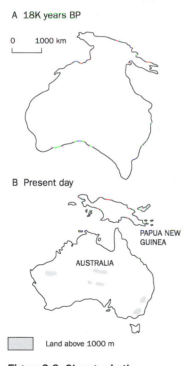

Figure 9.9 *Changes in the configuration of Australia's coastline. (A) 18K years BP, (B) At present*

Source: Based on Williams *et al.* (1993).

9.6 Conclusions

Low-latitude regions provide a wealth of information on environmental change over the last 3×10^6 years. The amount of information, from a wide variety of archives, will undoubtedly increase as new sites are investigated in a large and environmentally diverse region where there is a scarcity of research as compared with the high- and middle-latitude regions (Chapters 7 and 8). The highland zone of low latitudes provides information on direct glacial activity and, in common with the record from middle and high latitudes, it is evident that numerous glacial advances have occurred. Together with evidence from other archives of data on environmental change from areas not directly glaciated, the data from the highland zone attest to the dynamism of the environment during the last 3×10^6 years and thus run counter to traditional ideas of relative climatic and environmental stability within the low latitudes.

In particular, the unglaciated upland zone and the humid zone of the low latitudes have provided archives, mainly lake sediments and peats, with relatively long records of environmental change. The pollen assemblages from such archives provide information on the composition and distribution of past vegetation communities. On the basis of comparisons with the ecological requirements of present-day taxa in terms of temperature and precipitation ranges, inferences on past conditions can be made. Such studies have not only contributed to the determination of past environmental change, but have also contributed to important ecological debates. For example, the characteristics of plant communities spatially and temporally are now considered to be a culmination of the reaction of individual species to environmental change; they are not a reflection of the constancy of plant communities in terms of their composition, or a product of the same reaction of individual species to external stimuli such as climatic change. Palaeoecological research has also contributed to the high-lighting of inadequacies in the **refugia** hypothesis. The increased understanding of the genetic basis for speciation, in conjunction with research on past and present plant and animal distributions, is likely to contribute to the issue of high diversity in tropical regions.

Whilst research on ice cores from low latitudes is in its infancy, the Huascarán data (Box 9.3) illustrate the importance of such proxy climatic data for correlation with ice records elsewhere. The reconstruction of temperature change within a glacial–interglacial cycle, for example, is possible from oxygen isotope data. The Huascarán record in comparison with reconstructions from pollen data indicates that the temperature range was greater at high altitudes than in the lowlands. Confirmation of this trend requires further verification. However, there is a paucity of data from low-latitude arid and densely forested regions. Here, direct and proxy records of climatic change are elusive and, to date, the record is relatively sparse.

Summary Points

- The low latitudes are a rich, but relatively unexploited, source of palaeoenvironmental information.
- There are records of direct glaciations in the highlands of Africa, the Himalayas and the Andes.
- The non-glaciated upland zone and the lowland humid zone provide the most abundant archives of palaeoenvironmental data.
- Lake sediments and peats are the most abundant and widely investigated archives.

- Pollen analysis is a key technique in palaeoenvironmental reconstruction.

- The South American Andes have yielded highly significant information, notably from the long record of Funza, near Bogotá, Colombia, and the Huascarán ice core of Peru.

- All the evidence points to the low latitudes as being characterised by change rather than stability over the last 3×10^6 years.

- Palaeoenvironmental data from low latitudes are contributing to debates about how plant communities react to environmental change, the relevance of the **refugia** hypothesis, and the reason for high biotic diversity in the tropics.

General further reading

Environmental Change. A.S. Goudie. 1992. Clarendon Press, Oxford, 3rd edn.

Plant Community History: Long-Term Changes in Plant Distribution and Diversity. J.H. Tallis. 1991. Chapman and Hall, London.

Quaternary Environments. M.A.J. Williams, D.L. Dunkerley, P. de Deckker, A.P. Kershaw and T. Stokes. 1998. Arnold, London, 2nd edn.

Quaternary Geology and Geomorphology of South America. C.M. Clapperton. 1993. Elsevier, Amsterdam.

Reconstructing Quaternary Environments. J.J. Lowe and M.J.C. Walker. 1997. Longman, Harlow, Essex, 2nd edn.

 Conclusions

10.1 Preamble

The last 3×10^6 years were indeed dynamic in terms of environmental change. Moreover, the collection of data on environmental change, and ideas relating to it, have been and continue to be dynamic. The inauguration of the glacial theory in the mid-nineteenth century was a major turning point in the history of the earth and environmental sciences. In the 150 years since the theory was proposed, enormous strides have been made in data acquisition and analysis. Unsurprisingly, however, just as many new questions have been formulated as have been answered. The concerns, at the start of the new millennium, about possible global warming, have sharpened many of these questions and have injected a sense of urgency into palaeoenvironmental studies. This is because the past is the only available precedent for likely future change, although it must also be acknowledged that there may be no real precedents from any period in Earth history. The human-induced change characteristic of the present and the last 200 years or more is occurring at a rate that has no geological counterpart; human activity has accelerated the dynamism of all Earth-surface and atmospheric processes.

Nevertheless, the record of the last 3×10^6 years of environmental change provides many important 'lessons' that are directly relevant to future environmental change. First, the record provides a means of determining the relationships between various environmental systems, notably the oceans, atmosphere and continental environments. Second, it provides an opportunity to identify climatic/ecological thresholds, i.e. the identification of the magnitude of climatic change, notably of temperature and precipitation, which generates ecological change. Third, the record provides a means of identifying trends, their patterns and their periodicity. These are all vital components of planning for future environmental change.

10.2 The last 3×10^6 years: some generalisations

The geological record of the last 3×10^6 years, and earlier, provides evidence that astronomical **forcing factors**, i.e. the Milankovitch cycles (Box 2.2), are a key determinant of climatic change. These exhibit periodicities which are reflected in many archives of data on environmental change. During the last 3×10^6 years, these forcing factors have contributed to the initiation of a series of climatic cycles, each of which comprises a relatively long cold stage/ice age and a relatively short warm stage/**interglacial stage**. Each cycle lasts *c*. 120K years. The registration of these cycles in ocean sediments, ice sheets, and numerous continental archives, leaves little room for doubt that these factors are important in driving natural environmental change. These factors will continue to operate in the future, and they thus facilitate an element of prediction. If the sequence of ice ages/interglacials is to continue, another ice age might be expected to occur in *c*. 6K to 10K years.

However, the long-term geological history demonstrates that ice age conditions have not prevailed continuously throughout Earth history; they have been interspersed with long periods of quite different conditions (Figure 2.2). Some important questions, then, are what triggers the onset of ice ages, and what brings a succession of ice ages, as have been characteristic of the last 3×10^6 years, to a close? Undoubtedly, continental drift, mountain building, etc. have all played a role in these events. Other significant factors may include the Earth's biota and its reciprocal relationship with atmospheric composition. This is the essence of the **Gaia hypothesis**, which is discussed in Section 1.2.4. The polar ice cores, notably that from Vostok, Antarctica, which covers the last climatic cycle, have provided the first opportunity to determine directly the composition of past atmospheres through the analysis of air trapped in bubbles. Although it had already been considered a possibility, the resulting data (Figure 4.5) confirmed that substantial changes had occurred in the concentrations of both carbon dioxide and methane as each ice age ended and each interglacial began, and vice versa. Why and how this should occur, remain unresolved issues. Both carbon dioxide and methane are heat-trapping, i.e. greenhouse, gases, and are important components of the carbon **biogeochemical cycle**, linking the atmosphere, the biota and soils, and the oceans, through the processes of respiration, photosynthesis and decomposition. These processes link life with atmospheric composition, and hence with climatic change.

Hominid evolution from apes *c*. 5×10^6 to 7×10^6 years ago provides another twist to the tale. Eventually, *c*. 400×10^3 years ago, modern humans (*Homo sapiens sapiens*) evolved. By *c*. 10K years BP, *H. sapiens sapiens* was becoming a dominant factor within the carbon cycle, through the initiation and spread of agriculture. This transposed natural ecosystems into control systems and altered the dynamics of the carbon cycle. It also created wealth and facilitated technological development, which in turn led to industry and transport on a global scale.

These are attributes of human communities that have a substantial impact on the global carbon cycle. All are now recognised as contributing to global warming. Thus the reciprocity between life and environment, which is stressed by the Gaia hypothesis, continues. Humans, however, possess a far greater capacity to influence the carbon cycle than do other forms of life. Indeed, other forms of life are becoming extinct at an alarming rate because of this capacity. The rates of change in the carbon cycle, induced either deliberately or inadvertently by humans, have no earlier precedents. Whilst the results are, in general, unfavourable for other life forms, they may not in the context of the next few centuries, be all that favourable for humans either. Gaia will persist, as will life, but there are and will be extinctions. What is the prospect for humans? Moreover, have they altered atmospheric composition to such a degree already, with increased heat-trapping gas emissions in prospect, that the last 3×10^6 years of ice ages will be brought to an end, and another climatic regime begin? The lesson from the past record of natural environmental change is that there is a close link between the global carbon cycle and the global climate; the Earth must be managed wisely. To do this, the intricate operations of biogeochemical cycles must be identified and understood. The record of past environmental change should contribute to the identification and understanding of their components, with the observations and monitoring of modern processes playing an equally vital role in biogeochemical cycle management.

10.3 The last 3×10^6 years: some specifics

The record of environmental change over the last 3×10^6 years reflects a highly dynamic environment; the natural processes of change are ongoing and continual, in the sense that once a factor, such as climatic change, has occurred, the response will initiate a chain reaction within the biota and soils, etc. However, the response may not be immediate, or the culmination of the response, i.e. what is registered in the palaeoenvironmental archive, may be minor. In the case of palaeobotanical evidence, for example, most plant species survive within a range of annual temperatures, and, unless the range is exceeded as climatic change ensues, there will be little, if any, alteration in the record of pollen or macroscopic remains. In addition, other factors play a role within biotic communities: competition occurs between taxa, yet the impact of this on the palaeobotanical record is virtually impossible to determine. This reflects the imprecise nature of the methodology. Such shortcomings are also exemplified by the fact that the level of taxonomic resolution of pollen analysis (Box 5.1) – one of the most widely used techniques in palaeoenvironmental studies – is generally only possible to genus rather than taxon (species) level. Yet it is individual species, not whole families or indeed whole plant communities, that react to environmental change. Most palaeoenvironmental techniques have similar limitations. This is one of the reasons why multidisciplinary studies are particularly appropriate for palaeoenvironmental research.

Further limitations are placed on palaeoenvironmental data, especially in relation to the determination of rates of change, by age-estimation techniques. Either the techniques available have limitations, notably margins of error which curtail precision, or the period of time that a particular technique can address is restricted. Both problems are reflected by radiocarbon age estimation: ages are expressed in radiocarbon years with a margin of error, so that the greater the age of the sample the larger the error, meaning that, unfortunately, this technique is only applicable to material from the last 50K years. This is only c. 2.5 per cent of Quaternary time, and only 0.16 per cent of the last 3×10^6 years. All other available techniques of age estimation have similar limitations. However, perhaps the greatest drawback is the lack of a technique that has universal application, i.e. a technique that is viable for all of the last 3×10^6 years, or longer, and is applicable to a wide range of materials. This, of course, would be ideal, because it would facilitate direct comparison and correlation between disparate deposits. The inadequacies of age-estimation techniques, along with the technology and expense associated with them, preclude the determination of precise rates of environmental/ecological change. This also makes it difficult to identify the environmental/ecological thresholds in past ecological systems.

Despite these inadequacies, age-estimation techniques in general, and radiocarbon in particular, have made a substantial contribution to the elucidation of environmental change. For example, sequences deposited at the end of the last ice age, i.e. the late glacial period, which have numerous radiocarbon age estimations, are especially important, because they highlight the dynamism of environmental change and the rapidity with which it can occur. As stated above, the whole of the last 3×10^6 year period was characterised by a changing rather than uniform environment, which was largely driven by climatic change; the operation of c. 120K year climatic cycles is particularly apparent. However, the late glacial period, i.e. 14K to 10K years BP, was especially dynamic (Figure 4.10). Temperature changes of considerable magnitude were occurring, as exemplified by the situation in Britain, where, during the first warming phase at c. 13.5K years BP, temperature increases of almost 3°C per century were occurring (Box 8.2). Subsequently, rapid temperature depression occurred, i.e. during the **Younger Dryas** cooling event.

The widespread evidence, in terms of varied latitudes and altitudes, for these changes, indicate that this period was climatically complex on a global basis. The 'lesson' from these records is that climatic change is not necessarily a slow, long-term process; the global climatic system is capable of reacting to external and/or internal stimuli on time scales that include the centennial and decadal. Complacency about the rapidity with which anthropogenic global warming might occur is thus not warranted. In addition, the palaeoenvironmental data from the late glacial period reflect the ecological reaction to the climatic change; this too was dynamic on short time scales. Future climatic change then, whether natural or anthropogenic, is almost certainly going to have a major impact on the Earth's remaining natural ecosystems, including those of the oceans, and on its agricultural systems. In terms of future research, increasing the resolution of all types of data on this period is most important, along with the identification of the mechanisms and the role of the various factors involved. For example, how do Milankovitch factors relate to the global **biogeochemical cycle** of carbon? Further data on other equally dynamic periods, notably the opening and close of the cold/warm stages of the last 3×10^6 years, would also add to the understanding of particularly environmentally dynamic periods.

Climatic reconstructions from diverse parts of the world provide information on spatial variations. During the last ice age, for example, temperature depression was not globally uniform. Whilst generalisation is risky when comparing data sets collected in many different ways, there does appear to be a pattern: the high and middle latitudes and high altitudes, including those in the tropics, experienced a greater depression of temperature during the last ice age, and possibly previous ice ages, than low latitudes, and possibly low altitudes. During the ensuing warm stage, the Holocene, the rise in temperature was greatest in high and middle latitudes, and greater in high altitudes than in low latitudes. The 'lesson' is that the Earth does not cool or warm uniformly. It is interesting to note that **general circulation models** (GCMs) predict that the pattern and intensity of future anthropogenically induced global warming will also vary latitudinally, and is likely to follow a similar pattern to that described for natural climatic change.

Such concurrence increases confidence in GCMs. This reflects an important role for palaeoenvironmental data in the validation of GCMs. Their accuracy is paramount for future planning, but, because of their complexity, which in turn reflects the intricate relationships between the many components, they need to be tested rigorously. Palaeoenvironmental data provide a means of reconstructing past climatic characteristics and climatic scenarios at various points in time. GCMs can then be used to reconstruct climatic characteristics at the same points in time. This is a means of testing the GCMs; any discrepancies between the two can be investigated to determine causes, and adjustments can then be made to the GCMs if appropriate.

10.4 Envoi

The record of environmental change over the last 3×10^6 years is complex. Evidence from a variety of archives has facilitated the identification of overall trends or patterns, notably the operation of Milankovitch cycles, which drive global climatic change. The most pronounced of these patterns comprised cold stage/**glacial** to warm stage/**interglacial** shifts which were characterised by changes in average global temperatures of between 5°C and 10°C. Even this shift was not globally uniform, with the greatest temperature differences occurring in high and middle–high latitudes. There is also abundant evidence to show that climatic change occurred within each glacial and interglacial. Stability is, therefore, not a term that describes accurately the climates of the past 3×10^6 years. If there is a constant feature of this period, it is that the Earth's condition was generally colder than it is at present; the relatively short bursts of interglacial conditions could thus be described as 'out of character'.

Two additional intriguing considerations emerge from current understanding of the operation of these past climatic cycles. The first is that the Milankovitch forcing of climate, which causes glacial–interglacial cycles, is generally considered to be inadequate on its own to promote the necessary temperature changes. There appears to be a relationship with the global carbon cycle and the alteration of atmospheric composition in terms of carbon dioxide and methane concentrations. The mechanisms involved are imperfectly understood but, in true **Gaia** fashion, the Earth's **biota**, amongst other factors, is involved, through its role as a mode of flux within the cycle and as a means of carbon storage. The question is: how do Milankovitch cycles affect the carbon cycle? Possibly Milankovitch **forcing** lowers global temperatures sufficiently to generate a reaction within the global carbon cycle. This causes the release of carbon dioxide and methane into the atmosphere, thereby enhancing the greenhouse effect and thus intensifying global warming to bring about interglacial conditions. Milankovitch-forced cooling at the close of an interglacial must, therefore, stimulate carbon storage and the removal of carbon dioxide and methane from the atmosphere. Whatever the reality, the relationship between the carbon cycle and global temperatures over the last 3×10^6 years requires further elucidation. The second consideration raised by the recognition of climatic cycles is the relative rapidity with which glacial stages end and interglacials begin. This highlights the dynamism of the global climate and its capacity to react quickly, geologically speaking, to external and internal stimuli; changes in the global carbon cycle are examples of the latter. This capacity of the global climatic system to react quickly means that anthropogenic stimuli may bring about change more rapidly than society had realised was possible, and with unforeseen consequences.

The research referred to in this book reflects the widespread impact of climatic cycles and the climatic variability within them. It emphasises the wide variety of archives of palaeoenvironmental data that exist, and the variety of techniques that exist to reconstruct past environments. All have their drawbacks, and all are constrained by the limitations of age-estimation techniques. Nevertheless, palaeoecological and geological/ geomorphological data from diverse archives and regions, including the oceans, indicate that the Earth's surface biotic and abiotic components are as dynamic as its climatic system.

The most pronounced Earth-surface changes occurred, inevitably, in regions subject to direct glaciation. In these high and middle–high latitudes and high altitudes, an ice cover characteristic of glacial stages gave way to tundra, boreal or temperate forest vegetation communities during interglacials and often through the intermediate stages. The forests and steppes of today's mediterranean-type and Mediterranean and savannah lands witnessed plant-community displacement and disintegration as individual species reacted to climatic cooling and warming. Even in regions where the overall type of vegetation remained nearly constant, as in some tropical forests, there is evidence for change rather than constancy, in terms of species composition. Such information is valuable, not only in terms of its relevance to past environments, but also because it contributes to ecological theory through facilitating the testing of hypotheses based on conjecture, e.g. the community versus the individual species response to climatic change, and the validity of the concept of tropical forest **refugia**.

Finally, it is probable that past environmental change, caused by climatic change, to which it also contributed, facilitated the evolution of modern humans, who have since become a major driving force for environmental change, most notably through their impact on the global **biogeochemical cycle** of carbon. Discriminating between natural and anthropogenic environmental change is difficult, especially for the Holocene, and will become increasingly difficult as anthropogenic activity intensifies. The only certainty about environmental change is that it will continue at all spatial and temporal scales. Gaia is restless, vigorous and robust.

Summary Points

- Astronomical (Milankovitch) **forcing factors** generate cycles of climatic change to produce a degree of regularity within the geological record.

- There is a complex relationship between astronomical factors and the global **biogeochemical cycle** of carbon, which generates climatic and, in turn, environmental change.

- Climatic and environmental change can occur rapidly, on the centennial or possibly even the decadal scale.

- Methods of environmental reconstruction and age-determination techniques have their limitations.

- Past environmental change generated conditions suitable for the evolution and development of modern humans.

- The record of environmental change during the last 3×10^6 years lends support to the **Gaia hypothesis**.

- Records of past climatic change provide an important means of validating **general circulation models** (GCMs).

- Foci of further research include the detailed reconstruction of periods of abrupt environmental change, the role of biogeochemical cycling, especially of carbon, in climatic change, and the interrelationships between the oceans, atmosphere, and **biota**.

General further reading

Global Environmental Change. A Natural and Cultural Environmental History. A.M. Mannion. 1997. Longman, Harlow, 2nd edn.

Global Environmental Research (Swindon) Internet Home Page: http://www.nerc.ac.uk/ ukgeroff/welcome.htm

Reconstructing Quaternary Environments. J.J. Lowe and M.J.C. Walker. 1997. Longman, Harlow, 2nd edn.

The Globe 1998. *Special Issue on Past Global Change* **41** (available from the UK Global Environmental Research Office, David Phillips Building, North Star Avenue, Swindon SN2 1EU).

 **Glossary**

alkenones organic molecules present in the lipid (fat) components of cell membranes.

alkenone unsaturation index the degree of saturation or unsaturation of the alkenone content of marine organisms such as **coccolithophores** (see p. 42).

angiosperms the flowering plants.

benthic organisms organisms that inhabit the surface or near-surface mud of ocean/lake beds.

biogenic material remains of organisms and organic remains derived from the decomposition or fossilisation of organisms.

biogeochemical cycles the exchange of matter, i.e. carbon, nitrogen, hydrogen, etc., between the various components of the Earth's surface, including the atmosphere and living organisms.

biogeography the study and explanation of the past and present distribution of organisms.

biota a collective term for all living organisms.

catastrophism a pre-1860s theory that Earth-surface characteristics were created by cataclysmic events such as earthquakes, volcanic eruptions and floods.

chitin; chitinous a complex carbohydrate of which the outer coverings of insects and crustaceans are composed.

chronostratigraphy; chronostratigraphic units the subdivision of rock strata into units that correspond with intervals of geological time, the age of which has been determined.

chrysophytes a group of planktonic (i.e. free-swimming in the water body, see below) algae that have an outer covering of scales composed of silica.

continental drift a theory published in 1915 by Alfred Wegener; it espouses the view that the present-day configuration of the continents was created by the breakup of a single continent, e.g. Pangaea.

cordillera; cordilleran a mountain range; of mountain origin.

dendrochronology the establishment of age by counting tree rings.

diurnal temperature range (DTR) the range of temperatures that occurs in a given location over a 24 hour period.

dynamic equilibrium a state of balance between the components and processes within an environmental system.

ecosystem a unit comprising plants, animals, micro-organisms and their environment.

environmental determinism a concept invoking the overriding control of the environment, and especially climate, on human activity.

feedback loops may be positive or negative and comprise reactions to stimuli within environmental systems, which, in the case of the former, bring about change or, in the case of the latter, maintain the *status quo*.

forcing factors stimuli from within or without environmental systems, which generate positive feedback and thus effect environmental change.

Gaia hypothesis a proposal formulated by James Lovelock in the 1970s, which invokes a reciprocal self-regulating relationship between life and its environment through geological time.

general circulation models complex models comprising numerous mathematical equations derived from weather/climatic data; their objective is the prediction of climatic change.

geographical cycle (also known as the **cycle of erosion**) a theory proposed by William Morris Davis in the late 1800s, which invokes various stages in landscape development as erosion proceeds following uplift.

geoidal or **geodetic eustasy** sea-level changes caused by the shape of the Earth, which is a geoid rather than spherical.

glacial stage (ice age) a period of one or more ice advances, which is separated from an earlier or later glacial stage by a warm interval known as an interglacial. The two together comprise a climatic cycle characteristic of the last 1.8×10^6 years.

gneiss a coarse-grained metamorphic rock with bands of quartz and feldspar alternating with bands of mica.

gyttja a fine organic mud.

halophyte, halophytic any plant species that is tolerant of salt in the soil or in the air.

Heinrich events the deposition from icebergs of **ice-rafted debris** (see below), which becomes incorporated in ocean sediments as distinct layers. The term derives from the scientist, Hans Heinrich, who established that such debris was deposited episodically.

heterotrophic organisms organisms that do not have the capacity to make their own food, and are thus reliant on organisms, notably green plants, i.e. the **primary producers** (see below), that have this ability through photosynthesis.

hominid evolution the evolution of modern humans and their ancestors, from the apes.

ice-rafted debris material eroded from landmasses by glaciers and ice sheets which transport it to the coast; ice breaks off as icebergs, which transport the rock and sediment into the oceans (see **Heinrich events** above).

interglacial period an interval of increased warmth and of 10K to 20K years duration, which occurs between glacial stages.

interstadial period an interval of increased warmth but of relatively short duration (*c.* 2K years), which occurs within glacial stages between periods of major ice advance.

isostasy; isostatic change the balance that exists as the Earth's crust 'floats' on the underlying core; changes in the height of the continental surface of the Earth as ice accumulates during glacials or disperses during interglacials.

jet streams high-speed winds above 12 000 m in the atmosphere in temperate latitudes; they move in a wavy pattern from east to west.

loess a fine-grained, yellow sediment, which produces fertile soils; it originated as glacial and fluvioglacial materials that were wind-blown from source areas and deposited elsewhere.

maar lake a lake in the crater of an extinct volcano.

marine isotope stages see **oxygen isotope stages**.

nanoplankton tiny plankton less than 100 μm in diameter; they inhabit the water body of the oceans.

open system an environmental system with boundaries through which energy and matter can flow.

orogeny; orogenesis mountain building, including the deformation of rocks, metamorphism and uplift.

oxygen isotope stages (also known as **marine isotope stages**) units within ocean-sediment cores that are identified on the basis of the ratios of the isotopes of oxygen (O^{16}/O^{18}), and whose age has been determined; they are used as reference horizons.

pedogenesis the natural process of soil formation.

pelagic (muds) the environment of the open ocean waters; pelagic muds are deposited on the floor of the ocean deeps.

plankton; planktonic species organisms in lacustrine or marine environments that occupy open water and are subject to movement via currents and wave direction.

plate tectonics a concept advanced in the 1960s to explain the structure of the Earth's crust; this is composed of a number of large blocks, i.e. plates of solid rock, which 'float' on the molten rock below.

primary producers organisms, mainly the green plants, which can manufacture their own food through photosynthesis.

psammophyte, psammophytic any plant species that is tolerant of, or adapted to, living in a sand-dune environment.

radiometric age determination the use of the radioactive decay properties of some elements, e.g. carbon, to determine age.

refugium (singular), **refugia** (plural); also **refuge** isolated remnants of vegetation/animal communities that existed as extensive units under former climatic regimes.

reinforcing factors factors within an environmental system that react to initial stimuli to accentuate the change(s) originated by those stimuli.

sclerophyll woodland woodland in which species with hard, small, leathery leaves predominate; such characteristics reduce water loss.

speleothems cave deposits, e.g. stalactites and stalagmites, which are redeposited calcium carbonate.

stadial a period of major ice advance during a glacial stage.

stromatolite an accumulation of limestone produced by algae.

tectonic a term relating to the structural characteristics of the Earth's crust and the forces that influence it; see also **plate tectonics**.

termination the abrupt end of a glacial stage.

terrigenous inorganic material derived from the land.

travertine a pale limestone deposited from solution, often around springs.

troposphere the lowest layer of the Earth's atmosphere.

tufa a porous or cellular deposit of calcium carbonate often formed around springs in areas of chalk or limestone bedrock.

uniformitarianism an idea proposed in the 1800s, which advocated that modern processes operating on the Earth's surface are the same as those that occurred throughout geological time.

varve a layer of sediment comprising coarse and fine material deposited in a lake basin over one year; annual layers can be distinguished and may be used as an age-estimation technique.

Younger Dryas a distinct cold period following initial warming at the end of the last glacial period. It may be a global phenomenon.

xerophyte, **xerophytic** any plant species that is adapted to survival in drought conditions.

Bibliography

Aber, J.S. 1991. 'The glaciation of northeastern Kansas', *Boreas* **20**, 297–314.

Aguirre, M.L. and Whatley, R.C. 1995. 'Late Quaternary marginal marine deposits and paleoenvironments from northeastern Buenos Aires Province, Argentina – a review', *Quaternary Science Reviews* **14**, 223–254.

Alekseev, M.N. 1997. 'Paleogeography and geochronology in the Russian eastern Arctic during the second half of the Quaternary', *Quaternary International* **41–2**, 11–15.

Alexandre, A., Meunier, J.D., Lezine, A.M., Vincens, A. and Schwartz, D. 1997. 'Phytoliths: indicators of grassland dynamics during the late Holocene in intertropical Africa'. *Palaeogeography Palaeoclimatology Palaeoecology* **136**, 213–229.

Andersen, B.G. and Borns, H.W. 1994. *The Ice Age World*. Scandinavian University Press, Oslo, Copenhagen and Stockholm.

Anderson, P.M., Lozhkin, A.V., Belaya, B.V., Glushkova, O.Yu. and Brubaker, L.B. 1997. 'A lacustrine pollen record from near altitudinal forest limit, Upper Kolyma Region, northeastern Siberia'. *The Holocene* **7**, 331–335.

Anderson, T.W. and Macpherson, J.B. 1994. 'Wisconsinan late-glacial environmental change in Newfoundland: a regional review', *Journal of Quaternary Science* **9**, 171–178.

Andren, T. and Sohlenius, G. 1995. 'Late Quaternary development of the north western Baltic proper – results from the clay-varve investigation', *Quaternary International* **27**, 5–10.

Andrews, J.T. 1994. 'Wisconsinan late-glacial environmental change on the southeast Baffin Shelf, southeast Baffin Island and northern Labrador', *Journal of Quaternary Science* **9**, 179–183.

Andrews, J.T., Miller, G.H., Vincent, J.-S. and Shilts, W.W. 1986. 'Quaternary correlations in Arctic Canada', *Quaternary Science Reviews* **5**, 243–249.

Arkhipov, S.A., Bespaly, V.G., Faustova, M.A., Glushkova, O.Y., Isayeva, L.L. and Velichko, A.A. 1986a. 'Ice-sheet reconstructions', *Quaternary Science Reviews* **5**, 475–488.

Arkhipov, S.A., Isayeva, L.L., Bespaly, V.G. and Glushkova, O.Y. 1986b. 'Glaciation of Siberia and northeast USSR', *Quaternary Science Reviews* **5**, 463–474.

Atalay, I. 1996. 'Palaeosols as indicators of the climatic changes during the Quaternary period in S. Anatolia', *Journal of Arid Environments* **32**, 23–35.

Atkinson, T.C., Briffa, K.R. and Coope, G.R. 1987. 'Seasonal temperatures in Britain during the past 22,000 years, reconstructed using beetle remains', *Nature* **325**, 587–592.

Augustinus, P.C., Short, S.A. and Colhoun, E.A. 1994. 'Pleistocene stratigraphy of the Boco Plain, Western Tasmania', *Australian Journal of Earth Sciences* **41**, 581–591.

Baillie, M.G.L. 1995. *A Slice Through Time. Dendrochronology and Precision Dating.* B.T. Batsford, London.

Ballantyne, C.K. and Harris, C. 1994. *The Periglaciation of Great Britain.* Cambridge University Press, Cambridge.

Bard, E. and Broecker, W.S. (eds) 1992. *The Last Deglaciation: Absolute and Radiocarbon Chronologies.* Springer-Verlag, Berlin.

Bard, E., Jouannic, C., Hamelin, B., Pirazzoli, P., Arnold, M. and Faure, G. 1996. 'Pleistocene sea levels and tectonic uplift based on dating of corals from Sumba Island, Indonesia', *Geophysical Research Letters* **23**, 1473–1476.

Barnola, J.M., Raynaud, D., Korotkevich, Y.S. and Lorius, C. 1987. 'Vostok ice core provides 160,000-year record of atmospheric CO_2', *Nature* **329**, 408–414.

Barrett, P.J. 1991. 'Antarctica and global climatic change: a geological perspective', in *Antarctica and Global Climatic Change*, C. Harris and B. Stonehouse (eds). Belhaven Press, London, in association with the Scott Polar Research Institute, University of Cambridge, 35–50.

Barriendos, M. 1997. 'Climatic variations in the Iberian Peninsula during the late Maunder Minimum (AD 1675–1715): an analysis of data from rogation ceremonies', *The Holocene* **7**, 105–111.

Basile, I., Grousset, F.E., Revel, M., Petit, J.R., Biscaye, P.E. and Barkov, N.I. 1997. 'Patagonian origin of glacial dust deposited in East Antarctica (Vostok and Dome C) during glacial stages 2, 4 and 6', *Earth and Planetary Science Letters* **146**, 573–589.

Battarbee, R.W. 1994. 'Diatoms, lake acidification and the Surface Water Acidification Programme (SWAP) – a review', *Hydrobiologica* **274**, 1–7.

Baumann, K.H., Lackschewitz, K.S., Mangerud, J., Spielhagen, R.F., Wolfwelling, T.C.W., Heinrich, R. and Kassens, H. 1995. 'Reflection of Scandinavian ice sheet fluctuations in Norwegian sea sediments during the past 150,000 years', *Quaternary Research* **43**, 185–197.

Beaulieu, J.-L. and Reille, M. 1992. 'Pollen records of the last climatic cycles in the Devès volcano craters (Massif Central, France) II', Lac du Bouchet. 8th International Palynological Congress, Aix-en-Provence, Program and Abstracts, 33.

Becker, B. 1993. 'A 11,000-year German oak and pine dendrochronology for radiocarbon calibration', *Radiocarbon* **35**, 201–213.

Becker, B. and Kromer, B. 1993. 'The continental tree-ring record – absolute chronology, ^{14}C calibration and climatic change at 11Ka BP', *Palaeogeography Palaeoclimatology Palaeoecology* **103**, 67–71.

Benn, D.I. and Evans, D.J.A. 1997. *Glaciers and Glaciation.* Arnold, London.

Bennett, K.D. 1989. 'A provisional map of forest types for the British Isles 5,000 years ago', *Journal of Quaternary Science* **4**, 141–144.

Benson, L. and Thompson, R.S. 1987. 'The physical record of lakes in the Great Basin', in *The Geology of North America. Volume K-3. North America and Adjacent Oceans During the Last Deglaciation*, W.F. Ruddiman and H.E. Wright Jr (eds). Geological Society of America, Boulder, Colorado, 241–260.

Benxing, Z. and Rutter, N. 1998. 'On the problem of Quaternary glaciations, and the extent and patterns of Pleistocene ice cover in the Qinghai–Xizang (Tibet) Plateau', *Quaternary International* **45/46**, 109–122.

Berggren, W.A., Hilgen, F.J., Langereis, C.G., Kent, D.V., Obradovich, J.D., Raffi, I., Raymo, M.E. and Shackleton, N.J. 1995. 'Late Neogene chronology: new perspectives in high-resolution stratigraphy', *Bulletin of the Geological Society of America* **107**, 1272–1287.

Berglund, B.E. (ed.) 1986. *Handbook of Holocene Palaeoecology and Palaeohydrology*. John Wiley and Sons, Chichester.

Berglund, B.E., Birks, H.J.B., Ralska-Jasiewiczowa, M. and Wright, H.E. (eds) 1996a. *Palaeoecological Events during the Last 15000 Years. Regional Syntheses of Palaeoecological Studies of Lakes and Mires in Europe*. John Wiley and Sons, Chichester.

Berglund, B.E., Digerfeldt, G., Engelmark, R., Gaillard, M-J., Karlsson, S., Miller, U. and Risberg, J. 1996b. 'Sweden', in *Palaeoecological Events during the Last 15,000 years. Regional Syntheses of Palaeoecological Studies of Lakes and Mires in Europe*, B.E. Berglund, H.J.B. Birks, M. Ralska-Jasiewiczowa and H.E. Wright (eds). John Wiley and Sons, Chichester, 233–280.

Bergsten, H. 1994. 'A high resolution record of Late glacial and early Holocene marine sediments from southwestern Sweden: with special emphasis on environmental changes close to the Pleistocene–Holocene transition and the influence of freshwater from the Baltic basin', *Journal of Quaternary Science* **9**, 1–12.

Berner, R.A. 1994. '3Geocarb II: a revised model of atmospheric CO_2 over Phanerozoic time', *American Journal of Science* **291**, 56–91.

Billard, A. and Orombelli, G. 1986. 'Quaternary glaciations in the French and Italian piedmonts of the Alps', *Quaternary Science Reviews* **5**, 407–411.

Björck, S., Kromer, B., Johnsen, S., Bennike, O., Hammarlund, D., Lemdahl, G., Possnert, G., Rasmussen, T.L., Wohlfarth, B., Hammer, C.U. and Spurk, M. 1996. 'Synchronized terrestrial–atmospheric deglacial records around the North Atlantic', *Science* **274**, 1155–1160.

Bloemendal, J., Liu, X. and Rolph, T.C. 1995. 'Correlation of the magnetic susceptibility stratigraphy of Chinese loess and the marine oxygen isotope record', *Earth and Planetary Science Letters* **131**, 371–380.

Boardman, J. 1996. 'Paleosols', in *Past Glacial Environments. Sediments, Forms and Techniques*, J. Menzies (ed.). Butterworth Heinemann, Oxford, 301–314.

Bonnefille, R., Riollet, G., Buchet, G., Icole, M., Lafont, R. and Arnold, M.N.A. 1995. 'Glacial/interglacial record from intertropical Africa: high resolution pollen and carbon data at Rusaka, Burundi', *Quaternary Science Reviews* **14**, 917–936.

Borromei, A.M. 1995. 'Pollen analysis of a Holocenic peat in the Valley of Andorra, Tierra del Fuego, Argentina'. *Revista Chilena de Historia Natural* **68**, 311–319.

Boyle, E.A. 1990. 'Quaternary deepwater paleoceanography', *Science* **249**, 863–870.

Bradbury, J.P. 1997. 'A diatom record of climate and hydrology for the past 200K from Owens Lake, California with comparison to other Great Basin records', *Quaternary Science Reviews* **16**, 203–219.

Bradley, R.S. 1985. *Quaternary Palaeoclimatology*. Allen and Unwin, London.

Bradley, R.S. and Jones, P.D. 1993. '"Little ice age" summer temperature variations: their nature and relevance to recent global warming trends', *The Holocene* **3**, 367–376.

Brassell, S.C., Eglinton, G., Marlowe, I.T., Pflaumann, U. and Sarnthein, M. 1986. 'Molecular stratigraphy: a new tool for climatic assessment', *Nature* **320**, 129–133.

Briffa, K.R., Jones, P.D., Schweingruber, F.H., Shiyator, S.G. and Cook, E.R. 1995. 'Unusual twentieth-century summer warmth in a 1,000-year temperature record from Siberia', *Nature* **376**, 156–159.

Broecker, W.S. and Van Donk, J. 1970. 'Insolation changes, ice volumes and the [18]O record in deep-sea cores', *Reviews of Geophysics and Space Physics* **8**, 169–197.

Brook, E.J., Sowers, T. and Orchardo, J. 1996. 'Rapid variations in atmospheric methane concentration during the past 110,000 years', *Science* **273**, 1089–1091.

Case, R.A. and MacDonald, G.M. 1995. 'A dendroclimatic reconstruction of annual precipitation for the western Canadian prairies since AD 1505 from *Pinus flexilis* James', *Quaternary Research* **44**, 267–275.

Chapman, M.R., Shackleton, N.J., Zhao, M. and Eglinton, G. 1995. 'Faunal and alkenone reconstructions of subtropical North Atlantic surface hydrography and paleotemperature over the last 28k yr', *Paleoceanography* **11**, 343–357.

Chappell, J., Omura, A., Esat, T., McCulloch, M., Pandolfi, J., Ota, Y. and Pillans, B. 1995. 'Reconciliation of late Quaternary sea levels derived from coral terraces at Huon Peninsula with deep sea oxygen isotope records', *Earth and Planetary Science Letters* **141**, 227–236.

Chappellaz, J., Barnola, J.M., Raynaud, D., Korotkevich, Y.S. and Lorius, C. 1990. 'Ice-core record of atmospheric methane over the past 160,000 years', *Nature* **345**, 127–131.

Chlachula, J., Rutter, N.W. and Evans, M.E. 1997. 'A late Quaternary loess–paleosol record at Kurtak, Southern Siberia', *Canadian Journal of Earth Sciences* **34**, 679–686.

Clapperton, C.M. 1994. 'The Quaternary glaciation of Chile: a review', *Revista Chilena de Historia Natural* **67**, 369–383.

Clapperton, C.M., Sugden, D.E., Kaufman, D.S. and McCulloch, R.D. 1995. 'The last glaciation in central Magellan Strait, southernmost Chile', *Quaternary Research* **44**, 133–148.

CLIMAP Project Members 1976. 'The surface of ice-age Earth', *Science* **191**, 1131–1137.

CLIMAP Project Members. 1981. 'Seasonal reconstructions of the Earth's surface at the last glacial maximum', *Geological Society of America Map and Chart Series*, MC-36.

Cohen, A.L. and Tyson, P.D. 1995. 'Sea surface temperature fluctuations during the Holocene off the south coast of Africa – implications for terrestrial climate and rainfall', *The Holocene* **5**, 304–312.

Cole, K. 1982. 'Late Quaternary zonation of vegetation in the eastern Grand Canyon', *Science* **217**, 1142–1145.

Cole, K. 1985. 'Past rates of change, species richness, and a model of vegetational inertia in the Grand Canyon, Arizona', *American Naturalist* **125**, 289–303.

Colhoun, E.A., Van der Geer, G., Fitzsimons, S.J. and Heusser, L.E. 1994. 'Terrestrial and marine records of the last glaciation from Western Tasmania: do they agree?', *Quaternary Science Reviews* **13**, 293–300.

Colinvaux, P.A., De Oliveira, P.E., Moreno, J.E., Miller, M.C. and Bush, M.B. 1996. 'A long pollen record from lowland Amazonia: forest and cooling in glacial times', *Science* **274**, 85–88.

Colinvaux, P.A., Bush, M.B., Steinitz-Kannan, M. and Miller, M.C. 1997. 'Glacial and postglacial pollen records from the Ecuadorian Andes and Amazon', *Quaternary Research* **48**, 69–78.

Colman, S.M. plus 85 others. 1997. 'Preliminary results of the first scientific drilling on Lake Baikal, Buguldeika site, southeastern Siberia', *Quaternary International* **37**, 3–17.

Colman, S.M., Peck, J.A., Karabanov, E.P., Carter, S.J., Bradbury, J.P., King, J.W. and

Williams, D.F. 1995. 'Continental climate response to orbital forcing from biogenic silica records in Lake Baikal', *Nature* **378**, 769–771.

Cook, E.R. and Kairiukistis, L.A. (eds) 1990. *Methods of Dendrochronology*. Kluwer, Dordrecht.

Cowie, J. W. and Bassett, M. G. 1989. 'Global stratigraphic chart with geochronometric and magnetostratigraphic calibration', *Episodes* **12**(2), Supplement.

Crane, P. R., Friis, S. M. and Pederson, K. R. 1995. 'The origin and early diversification of angiosperms', *Nature* **374**, 27–33.

Crombie, M.K., Arvidson, R.E., Sturchio, N.C., El Alfy, Z. and Abu Zeid, K. 1997. 'Age and isotopic constraints on Pleistocene pluvial episodes in the Western Desert, Egypt', *Palaeogeography Palaeoclimatology Palaeoecology* **130**, 337–355.

Cronin, T.M., Holtz, T.R. and Whatley, R.C. 1994. 'Quaternary paleoceanography of the deep Arctic Ocean based on quantitative analysis of Ostracoda', *Marine Geology* **119**, 305–332.

Cuffey, K.M., Clow, G.D., Alley, R.B., Stuiver, M., Waddington, E.D. and Saltus, R.W. 1995. 'Large Arctic temperature change at the Wisconsin–Holocene glacial transition', *Science* **270**, 455–458.

Currie, R.G. 1995. 'Luni-solar 18.6 and solar cycle 10–11 year signal in Chinese dryness–wetness indices', *International Journal of Climatology* **15**, 497–515.

Cwynar, L.C., Levesque, A.J., Mayle, F.E. and Walker, I. 1994. 'Wisconsinan late-glacial environmental change in New Brunswick: a regional synthesis', *Journal of Quaternary Science* **9**, 161–164.

Davis, W. L. and McKay, C. P. 1996. 'Origins of life – a comparison of theories and applications to Mars', *Origins of Life and Evolution of the Biosphere* **26**, 61–73.

Dawson, A.G. 1992. *Ice Age Earth*. Routledge, London.

De Angelis, M., Barkov, N.I. and Petrov, V.N. 1987. 'Aerosol concentrations over the last climatic cycle (160K yr) from an Antarctic ice core', *Nature* **325**, 318–321.

Delcourt, H.R. and Delcourt, P.A. 1991. *Quaternary Ecology. A Palaeoecological Perspective*. Chapman and Hall, London.

Demenocal, P.B. 1995. 'PlioPleistocene African climate', *Science* **270**, 53–59.

Derbyshire, E. 1996. 'Quaternary glacial sediments, glaciation style, climate and uplift in the Karakoram and northwest Himalaya: review and speculations', *Palaeogeography Palaeoclimatology Palaeoecology* **120**, 147–157.

Ding, Z., Hu, N., Rutter, N.W. and Liu, T. 1994. 'Towards an orbital time scale for Chinese loess deposits', *Quaternary Science Reviews* **13**, 39–70.

Dodonov, A.E. and Baiguzina, L.L. 1995. 'Loess stratigraphy of central Asia – palaeoclimatic and palaeoenvironmental aspects', *Quaternary Science Reviews* **14**, 707–720.

Donner, J. 1995. *The Quaternary History of Scandinavia*. Cambridge University Press, Cambridge.

Duk-Rodkin, A., Barendregt, R.W., Tarnocai, C. and Phillips, F.M. 1996. 'Late Tertiary to Late Quaternary record in the Mackenzie Mountains, Northwest Territories, Canada – stratigraphy, paleosols, paleomagnetism, and Cl-36', *Canadian Journal of Earth Sciences* **33**, 875–895.

Dupont, L.M. 1993. 'Vegetation zones in NW Africa during the Brunhes chron reconstructed from marine palynological data', *Quaternary Science Reviews* **12**, 189–202.

Dyke, A.S., Dale, J.E. and McNeely, R.N. 1996. 'Marine mollusks as indicators of environmental change in glaciated North America and Greenland during the last 18,000 years', *Géographie Physique et Quaternaire* **50**, 125–184.

Easterling, D.R., Horton, B., Jones, P.D., Peterson, T.C., Karl, T.R., Parker, D.E., Salinger, M.J., Razuvayev, V., Plummer, N., Jamason, P. and Folland, C.K. 1997. 'Maximum and minimum temperature trends for the globe', *Science* **277**, 364–367.

Eide, L.K., Beyer, J.K. and Jansen, E. 1996. 'Comparison of Quaternary interglacial periods in the Iceland Sea', *Journal of Quaternary Science* **11**, 115–124.

Elenga, H., Schwartz, D. and Vincens, A. 1994. 'Pollen evidence of late Quaternary vegetation and inferred climate changes in Congo', *Palaeogeography Palaeoclimatology Palaeoecology* **109**, 345–356.

Emeis, K.-C., Anderson, D.M., Doose, H., Kroon, D. and Schulz-Bull, D. 1995. 'Sea-surface temperatures and the history of monsoon upwelling in the northwest Arabian sea during the last 500,000 years', *Quaternary Research* **43**, 355–361.

Emiliani, C. 1955. 'Pleistocene temperatures', *Journal of Geology* **63**, 538–575.

Emiliani, C. 1995. '2 revolutions in the earth sciences', *Terra Nova* **7**, 587–597.

Evans, D. A., Beukes, D. A. and Kirschvink, J. L. 1997. 'Low latitude glaciation in the Palaeoproterozoic era', *Nature* **386**, 262–266.

Evison, L.H., Calkin, P.E. and Ellis, J.M. 1996. 'Late-Holocene glaciation and 20th century retreat, northeastern Brooks Range, Alaska', *The Holocene* **6**, 17–24.

Feng, Z.D. 1997. 'Geochemical characteristics of a loess–soil sequence in central Kansas', *Soil Science Society of America Journal* **61**, 534–541.

Ferguson, C.W. and Graybill, D.A. 1983. 'Dendrochronology of bristlecone pine: a progress report', *Radiocarbon* **25**, 287–288.

Fitzsimons, S.J. 1997. 'Late-glacial and early Holocene glacier activity in the Southern Alps, New Zealand', *Quaternary International* **38–9**, 69–76.

Flores, J.A., Sierro, F.J., Frances, G., Vazques, A. and Zamarreno, I.N.A. 1997. 'The last 100,000 years in the western Mediterranean: sea surface water and frontal dynamics as revealed by coccolithophores', *Marine Micropaleontology* **29**, 351–366.

Follieri, M., Magri, D. and Sadori, L. 1988. '250,000-year record from Valle di Castiglione (Roma)', *Quaternary International* **3/4**, 81–84.

French, H.M. 1996. *The Periglacial Environment.* Longman, Harlow, Essex, 2nd edn.

Fuji, N. 1988. 'Palaeovegetation and palaeoclimate changes around Lake Biwa, Japan during the last ca. 3 million years', *Quaternary Science Reviews* **7**, 21–28.

Fuji, N. and Horowitz, A. 1989. 'Brunhes epoch paleoclimates of Japan and Israel', *Palaeogeography Palaeoclimatology Palaeoecology* **72**, 79–88.

Funnell, B.M. 1996. 'PlioPleistocene palaeogeography of the southern North Sea basin', *Quaternary Science Reviews* **15**, 391–405.

Gale, S.J. and Hoare, P.G. 1997. 'The glacial history of the northwest Picos de Europa of northern Spain', *Zeitschrift für Geomorpholgie* **41**, 81–96.

Gellatly, A.F., Grove, J.M., Bucher, A., Latham, R. and Whally, W.B. 1994. 'Recent historical fluctuations of the glacier du Taillon, Pyrenees', *Physical Geography* **15**, 399–413.

Godwin, H. 1975. *The History of the British Flora: A Factual Basis for Phytogeography.* Cambridge University Press, Cambridge.

Goede, A., McDermott, F., Hawkesworth, C., Webb, J. and Finlayson, B. 1996. 'Evidence of Younger Dryas and neoglacial cooling in a late Quaternary paleotemperature record from a speleothem in Eastern Victoria, Australia', *Journal of Quaternary Science* **11**, 1–7.

Grootes, P.M., Stuiver, M., White, J.W.C., Johnsen, S. and Jouzel, J. 1993. 'Comparison of oxygen isotope records from the GISP2 and GRIP Greenland ice cores', *Nature* **366**, 552–554.

Grove, J.M. 1988. *The Little Ice Age*. Methuen, London.

Grove, J.M. and Conterio, A. 1995. 'The climate of Crete in the sixteenth and seventeenth centuries', *Climatic Change* **30**, 223–247.

Gupta, S.M. and Fernandes, A.A. 1997. 'Quaternary radiolarian faunal changes in the tropical Indian Ocean: inferences to paleomonsoonal oscillation of the 10 degrees S hydrographic front', *Current Science* **72**, 965–972.

Haffer, J. 1969. 'Speciation in Amazonian forest bird', *Science* **165**, 131–137.

Hahne, J. and Malles, M. 1997. 'Late- and post-glacial vegetation and climate history of the southwestern Taymyr Peninsula, central Siberia, as revealed by pollen analysis of a core from Lake Lama', *Vegetation History and Archaeobotany* **6**, 1–8.

Hald, M. and Aspeli, R. 1997. 'Rapid climatic shifts of the northern Norwegian Sea during the last deglaciation and the Holocene', *Boreas* **26**, 15–28.

Hall, B.L., Denton, G.H., Lux, D.R. and Schluchter, C. 1997. 'Pliocene palaeoenvironment and Antarctic ice sheet behaviour. Evidence from Wright Valley', *Journal of Geology* **105**, 285–294.

Hamilton, T.D. 1986. 'Correlation of Quaternary glacial deposits in Alaska', *Quaternary Science Reviews* **5**, 171–180.

Hansen, B.C.S. and Engstrom, D.R. 1996. 'Vegetation history of Pleasant Island, southeastern Alaska, since 13,000 year BP', *Quaternary Research* **46**, 161–175.

Harle, K.J. 1997. 'Late Quaternary vegetation and climate change in southeastern Australia: palynological evidence from marine core E55–6', *Palaeogeography Palaeoclimatology Palaeoecology* **131**, 465–483.

Harrison, S.P. and Dodson, J. 1993. 'Climates of Australia and New Guinea since 18,000 yr BP', in *Global Climates since the Last Glacial Maximum*, H.E. Wright Jr (ed.). University of Minnesota Press, Minneapolis, 265–292.

Hays, J.D., Imbrie, J. and Shackleton, N.J. 1976. 'Variations in the Earth's orbit: pacemaker of the ice ages', *Science* **194**, 1121–1132.

Hebbeln, D. and Wefer, G. 1997. 'Late Quaternary paleoceanography in the Fram Strait', *Paleoceanography* **12**, 65–78.

Heinrich, H. 1988. 'Origin and consequences of cyclic ice rafting in the northeast Atlantic Ocean during the past 130,000 years', *Quaternary Research* **29**, 142–152.

Heusser, C.J. 1993. 'Late-glacial of southern South America', *Quaternary Science Reviews* **12**, 345–350.

Heusser, L. and Morley, J. 1997. 'Monsoon fluctuations over the past 350K yr: high resolution evidence from northeast Asia northwest Pacific climate proxies (marine pollen and radiolarians)', *Quaternary Science Reviews* **16**, 565–581.

Hickman, M. and Schweger, C.E. 1996. 'The Late Quaternary paleoenvironmental history of a presently deep fresh-water lake in east central Alberta, Canada and paleoclimate implications', *Palaeogeography Palaeoclimatology Palaeoecology* **123**, 161–178.

Hjort, C., Ingolfsson, O., Moller, P., and Lirio, J.M. 1997. 'Holocene glacial history and sea-level changes on James Ross Island, Antarctic Peninsula', *Journal of Quaternary Science* **12**, 259–273.

Holmden, C., Geaser, R.A. and Muehlenbachs, K. 1997. 'Paleosalinities in ancient brackish water systems determined by Sr^{87}/Sr^{86} ratios in carbonate fossils: a case study from the Western Canada sedimentary basin', *Geochimica et Cosmochimica Acta* **61**, 2105–2118.

Hong, S.M., Candelone, J.P., Turetta, C. and Boutron, C.F. 1996. 'Changes in natural lead, copper, zinc and cadmium concentrations in central Greenland ice from 8250 to 149,100 years ago – their association with climatic changes and resultant variations of dominant source contributions', *Earth and Planetary Science Letters* **143**, 233–244.

Hooghiemstra, H. 1989. 'Quaternary and upper Pliocene glaciations and forest development in the tropical Andes: evidence from a long high resolution pollen record from the sedimentary basin of Bogotá, Colombia', *Palaeogeography Palaeoclimatology Palaeoecology* **72**, 11–26.

Hooghiemstra, H., Melici, J.L., Berger, A. and Shackleton, N.J. 1993. 'Frequency spectra and palaeoclimatic variability of the high resolution 30–1450 Ka Funza I pollen record (Eastern Cordillera, Colombia)', *Quaternary Science Reviews* **12**, 141–161.

Hope, G. and Tulip, J. 1994. 'A long vegetation history from lowland Irian-Jaya, Indonesia', *Palaeogeography Palaeoclimatology Palaeoecology* **109**, 385–398.

Horowitz, A. 1989. 'Continuous pollen diagrams for the last 3.5 M.Y. from Israel: vegetation, climate and correlation with the oxygen isotope record', *Palaeogeography Palaeoclimatology Palaeoecology* **72**, 63–78.

Hughes, M.K., Wu, X.D., Shao, X.M. and Garfin, G.M. 1994. 'A preliminary reconstruction of rainfall in north-central China since AD 1600 from tree-ring density and width', *Quaternary Research* **42**, 88–99.

Imbrie, J. and Imbrie, K.P. 1979. *Ice Ages. Solving the Mystery*. Macmillan, Basingstoke.

Imbrie, J., Berger, A. and Shackleton, N.J. 1993. 'Role of orbital forcing: a two-million-year perspective', in *Global Changes in the Perspective of the Past*, J.A. Eddy and H. Oeschger (eds). John Wiley and Sons, Chichester, 263–277.

Islebe, G.A. and Hooghiemstra, H. 1997. 'Vegetation and climate history of montane Costa Rica since the last glacial', *Quaternary Science Reviews* **16**, 589–604.

Ivanovich, M. and Harmon, R.S. (eds) 1995. *Uranium-series Disequilibrium: Applications to Earth, Marine and Environmental Sciences*. 2nd edn. Clarendon Press, Oxford.

Jacoby, G.C., D'Arrigo, R.D. and Davaajamts, T. 1996. 'Mongolian tree rings and 20th-century warming', *Science* **273**, 771–773.

Jenkins, D.G. 1987. 'Was the Plio-Pleistocene boundary placed at the wrong stratigraphic level?', *Quaternary Science Reviews* **6**, 41–42.

Jennings, S.A. 1995. Late Quaternary changes in pinyon pine and juniper distribution in the White Mountain region of California and Nevada. *Physical Geography* **16**, 276–288.

Jerardino, A. 1995. 'Late Holocene neoglacial episodes in southern South America and southern Africa – a comparison', *The Holocene* **5**, 361–368.

Jiang, J.M., Zhang, D.E. and Fraedrict, K. 1997. 'Historic climate variability of wetness in East China (960–1992): A wavelet analysis', *International Journal of Climatology* **17**, 969–981.

Johnsen, S.J., Clausen, H.B., Dansgaard, W., Fuhrer, K., Gundestrup, N., Hammer, C.U., Iversen, B., Jouzel, J., Stauffer, B. and Steffensen, J.P. 1992. 'Irregular glacial interstadials recorded in a new Greenland ice core', *Nature* **359**, 311–313.

Johnson, W.H., Hansel, A.K., Bettis, E.A., Karrow, P.F., Larson, G.J., Lowell, T.V. and

Schneider, A.F. 1997. 'Late Quaternary temporal and event classifications, Great Lakes region, North America', *Quaternary Research* **47**, 1–12.

Jolly, D., Taylor, D., Marchant, R., Hamilton, A., Bonnefille, R., Buchet, G. and Riollet, G. 1997. 'Vegetation dynamics in central Africa since 18,000 yr BP: pollen records from the interlacustrine highlands of Burundi, Rwanda and western Uganda', *Journal of Biogeography* **24**, 495–512.

Jones, A.M. 1997. '*Environmental Biology*', Routledge, London.

Jones, P.D., Briffa, K.R. and Schweingruber, F.H. 1995. 'Tree-ring evidence of the widespread effects of explosive volcanic eruptions', *Geophysical Research Letters* **22**, 1333–1336.

Jones, R.L. and Keen, D.H. 1993. *Pleistocene Environments in the British Isles*. Chapman and Hall, London.

Joseph, L., Moritz, C. and Hugall, A. 1995. 'Molecular support for vicariance as a source of diversity in rainforest', *Proceedings of the Royal Society of London* **B260**, 177–182.

Jouzel, J., Lorius, C., Petit, J.-R., Genthon, C., Barkov, N.I., Kotlyakov, V.M. and Petrov, V.M. 1987. 'Vostok ice core: a continuous isotope temperature record over the last climatic cycle (160,000 years)', *Nature* **329**, 403–408.

Jouzel, J., Waelbroeck, C., Malaize, B., Bender, M., Petit, J.R., Stievenard, M., Barkov, N.I., Barnola, J.M., King, T., Kotlyakov, V.M., Lipenkov, V., Lorius, C., Raynaud, D., Ritz, C. and Sowers, T. 1996. 'Climatic interpretation of the recently extended Vostok ice records', *Climate Dynamics* **12**, 513–521.

Kapsner, W.R., Alley, R.B., Shuman, C.A., Anandakrishnan, S. and Grootes, P.M. 1995. 'Dominant influence of atmospheric circulation in Greenland over the past 18,000 years', *Nature* **373**, 52–54.

Keigwin, L.D., Jones, G.A. and Lehman, S.J. 1991. 'Deglacial meltwater discharge, North Atlantic deep circulation, and abrupt climatic change', *Journal of Geophysical Research* **96**, 16811–16826.

Kershaw, A.P. 1986. 'The last two glacial–interglacial cycles from northeast Queensland: implications for climatic change and Aboriginal burning', *Nature* **322**, 47–49.

Kershaw, A.P. 1994. 'Pleistocene vegetation of the humid tropics of northeast Queensland, Australia', *Palaeogeography Palaeoclimatology Palaeoecology* **109**, 399–412.

Krzyskowski, D., Bottger, T., Junge, F.W., Kuszell, T. and Nawrocki, J. 1996. 'Ferdynandovian interglacial climate reconstructions from pollen successions, stable-isotope composition and magnetic susceptibility', *Boreas* **25**, 283–296.

Kukla, G., McManus, J.F., Rousseau, D.-D. and Chuine, I. 1997. 'How long and how stable was the last interglacial?', *Quaternary Science Reviews* **16**, 605–612.

Kuniholm, P.I., Kromer, B., Manning, S.W., Newton, M., Latini, C. and Bruce, M.J. 1996. 'Anatolian tree rings and the absolute chronology of the eastern Mediterranean', *Nature* **381**, 780–783.

Lamb, H.H. 1995. *Climate, History and the Modern World*. Routledge, London, 2nd edn.

Lara, A. and Villalba, R. 1993. 'A 3620-year temperature record from *Fitzroya cupressoides* tree rings in southern South America', *Science* **260**, 1104–1106.

Lauritzen, S.E. 1995. 'High resolution paleotemperature proxy record for the last interglaciation based on Norwegian speleothems', *Quaternary Research* **43**, 133–146.

Lebreiro, S.M., Moreno, J.C., McCave, I.N. and Weaver, P.P.E. 1996. 'Evidence for Heinrich layers off Portugal (Tore seamount 39 degrees N, 12 degrees W)', *Marine Geology* **131**, 47–56.

Legrand, M. 1997. 'Ice-core records of atmospheric sulphur', *Philosophical Transactions of the Royal Society of London Series B* **352**, 241–250.

Legrand, M.R., Lorius, C., Barkov, N.I. and Petrov, V.N. 1988. 'Vostok (Antarctica) ice core: atmospheric chemistry changes over the last climate cycle (160,000 years)', *Atmospheric Environment* **22**, 317–331.

Leroy, S. and Dupont, L.M. 1994. 'Development of vegetation and continental aridity in northwestern Africa during the Late Pliocene: the pollen record of ODP Site 658', *Palaeogeography Palaeoclimatology Palaeoecology* **109**, 295–316.

Licht, K.J., Jennings, A.E., Andrews, J.T. and Williams, K.M. 1996. 'Chronology of Late Wisconsin ice retreat from the Western Ross Sea, Antarctica', *Geology* **24**, 223–226.

Lorius, C. and Oeschger, H. 1994. 'Palaeo-perspectives: reducing uncertainties in global change?', *Ambio* **23**, 30–36.

Lorius, C., Jouzel, J., Ritz, C., Merlivat, L., Barkov, N.I., Korotkevich, Y.S. and Kotlyakov, V.M. 1985. 'A 150,000 year climatic record from Antarctic ice', *Nature* **316**, 591–596.

Lovelock, J. 1995. *The Ages of Gaia. A Biography of Our Living Earth*. Oxford University Press, Oxford, 2nd edn.

Lovelock, J. 1997. 'A geophysiologist's thoughts on the natural sulphur cycle', *Philosophical Transactions of the Royal Society of London* **B352**, 143–147.

Lowe, J.J. and Walker, M.J.C. 1997. *Reconstructing Quaternary Environments*. Longman, Harlow, Essex, 2nd edn.

Lowe, J.J., Coope, G.R., Sheldrick, C., Harkness, D.D. and Walker, M.J.C. 1995. 'Direct comparison of UK temperatures and Greenland snow accumulation rates, 15,000–12,000 years ago', *Journal of Quaternary Science* **10**, 175–180.

Lozhkin, A.V. and Anderson, P.M. 1995. 'The last interglaciation in Northeast Siberia', *Quaternary Research* **43**, 147–158.

McManus, J.F., Bond, G.C., Broecker, W.S., Johnsen, S., Labeyrie, L. and Higgins, S. 1994. 'High resolution climate records from the North Atlantic during the last interglacial', *Nature* **371**, 326–329.

Magee, J.W., Bowler, J.M., Miller, G.H. and Williams, D.L.G. 1995. 'Stratigraphy, sedimentology, chronology and paleohydrology of Quaternary lacustrine deposits at Madigan Gulf, Lake Eyre, South Australia', *Palaeogeography Palaeoclimatology Palaeoecology* **113**, 3–42.

Mahaney, W.C. 1990. *Ice on the Equator. Quaternary Geology of Mount Kenya, East Africa*. Wm. Caxton, Sister Bay, Wisconsin, USA.

Mannion, A.M. 1995. *Agriculture and Environmental Change. Temporal and Spatial Dimensions*. John Wiley and Sons, Chichester.

Mannion, A. M. 1997a. 'Climate and vegetation', in *Applied Climatology. Principles and Practice*, R.D. Thompson and A. Perry (eds). Routledge, London, 123–140.

Mannion, A.M. 1997b. *Global Environmental Change. A Natural and Cultural Environmental History*. Longman, Harlow, Essex, 2nd edn.

Markgraf, V. 1993. 'Younger Dryas in southernmost South America – an update', *Quaternary Science Reviews* **12**, 351–355.

Martin, J.H. 1990. 'Glacial-interglacial CO_2 change: the iron hypothesis', *Paleoceanography* **5**, 1–13.

Martin, J.H. plus 43 others. 1994. 'Testing the iron hypothesis in ecosystems of the equatorial Pacific Ocean', *Nature* **371**, 123–129.

Mason, O.K., Salmon, D.K. and Ludwig, S.L. 1996. 'The periodicity of storm surges in the Bering Sea from 1898 to 1993, based on newspaper accounts', *Climatic Change* **34**, 109–123.

Metcalfe, S.E., Bimpson, A., Courtice, A.J., O'Hara, S.L. and Taylor, D.M. 1997. 'Climate change at the monsoon/westerly boundary in Northern Mexico', *Journal of Paleolimnology* **17**, 155–171.

Meyers, P.A. and Takemura, K. 1997. 'Quaternary changes in delivery and accumulation of organic matter in sediments of Lake Biwa, Japan', *Journal of Paleolimnology* **18**, 211–218.

Moe, D., Vorren, K.-D., Alm, T., Fimreite, S., Mørkved, B., Nilssen, E., Paus, A.A. Ramfjord, H., Selvik, S.F. and Sørensen, R. 1996. 'Norway', in *Palaeoecological Events During the Last 15000 years. Regional Syntheses of Palaeoecological Studies of Lakes and Mires in Europe*, B.E. Berglund, H.J.B. Birks, M. Ralska-Jasiewiczowa and H.E. Wright (eds). John Wiley and Sons, Chichester, 153–213.

Mommersteeg, H.J.P.M., Loutre, M.F., Young, R., Wijmstra, T.A. and Hooghiemstra, H. 1995. 'Orbital forced frequencies in the 975000-year pollen record from Tenaghi–Philippon (Greece)', *Climate Dynamics* **11**, 4–24.

Moore, P.D., Webb, J.A. and Collinson, M.D. 1991. *Pollen Analysis*. Blackwell, Oxford, 2nd edn.

Morley, J. J. and Dworetzky, B. A. 1991. 'Evolving Pliocene–Pleistocene climate: a North Pacific perspective', *Quaternary Science Reviews* **10**, 225–238.

Mott, R.J., 1994. 'Wisconsinan late-glacial environmental change in Nova Scotia: a regional synthesis', *Journal of Quaternary Science* **9**, 155–160.

Neumann, K. 1991. 'In search of the green Sahara: palynology and botanical macro-remains', *Palaeoecology of Africa* **22**, 203–212.

Nilsson, T. 1983. *The Pleistocene. Geology and Life in the Quaternary Ice Age*. Reidel, Dordrecht.

Ning, S. and Dupont, L.M. 1997. 'Vegetation and climatic history of southwest Africa: a marine palynological record of the last 300,000 years', *Vegetation History and Archaeobotany* **6**, 117–131.

O'Brien, S.R., Mayewski, P.A., Meeker, L.D., Meese, D.A., Twickler, M.S. and Whitlow, S.I. 1995. 'Complexity of Holocene climate as reconstructed from a Greenland ice core', *Science* **270**, 1962–1964.

Oeschger, H. and Langway, C.C. Jr (eds) 1989. *The Environmental Record in Glaciers and Ice Sheets*. John Wiley and Sons, New York.

Overpeck, J., Hughen, K., Hardy, D., Bradley, R., Case, R., Douglas, M., Finney, B., Gajewski, K., Jacoby, G., Jennings, A., Lamoureux, S., Lasca, A., MacDonald, G., Moore, J., Retelle, M., Smith, S., Wolfe, A. and Zielinski, G. 1997. 'Arctic environmental change of the last four centuries', *Science* **278**, 1251–1256.

Oviatt, C.G. 1997. 'Lake Bonneville fluctuations and global climate change', *Geology* **25**, 155–158.

Parker, D.E., Wilson, H., Jones, P.D., Christy, J.R. and Folland, C.K. 1996. 'The impact of Mount Pinatubo on world-wide temperatures', *International Journal of Climatology* **16**, 487–497.

Partridge, T.C. 1997. 'Cainozoic environmental change in southern Africa, with special emphasis on the last 200,000 years', *Progress in Physical Geography* **21**, 3–22.

Peteet, D.M., Darvels, R.A., Heusser, L.E., Vogel, J.S., Southon, J.R. and Nelson, D.E.

1994. 'Wisconsinan late-glacial environmental change in southern New England: a regional synthesis', *Journal of Quaternary Science* **9**, 151–154.

Pfister, C. 1992. 'Monthly temperature and precipitation in central Europe 1525–1979: quantifying documentary evidence on weather and its effects', in *Climate Since AD 1500*, R.S. Bradley and P.D. Jones (eds). Routledge, London (reprinted with revisions in 1995), 118–142.

Pfister, C., Schwarzzanetti, G. and Wegmann, M. 1996. 'Winter severity in Europe – the 14th century', *Climatic Change*, **34**, 91–108.

Phillips, R.L. and Grantz, A. 1997. 'Quaternary history of sea ice and paleoclimate in the Amerasia basin, Arctic Ocean, as recorded in the cyclical strata of Northwind Ridge', *Geological Society of America Bulletin* **109**, 1101–1115.

Pilcher, J.R., Baillie, M.G.L., Schmidt, B. and Becker, B. 1984. 'A 7272-year tree-ring chronology for western Europe', *Nature* **312**, 150–152.

Pillans, B. 1991. 'New Zealand Quaternary stratigraphy: an overview', *Quaternary Science Reviews* **10**, 405–418.

Pillans, B. 1994. 'Direct marine terrestrial correlations, Wanganui Basin, NE New Zealand – the last 1 million years', *Quaternary Science Reviews* **13**, 189–200.

Pisias, N.G. and Mix, A.C. 1997. 'Spatial and temporal oceanographic variability of the eastern equatorial Pacific during the late Pleistocene: evidence from radiolaria microfossils', *Paleoceanography* **12**, 381–393.

Pisias, N.G., Roelofs, A. and Weber, M. 1997. 'Radiolarian based transfer functions for estimating mean surface ocean temperatures and seasonal change', *Paleoceanography* **12**, 365–379.

Pollock, D.E. 1997. 'The role of diatoms, dissolved silicate and Antarctic glaciation in glacial/interglacial climatic change: a hypothesis', *Global and Planetary Change* **14**, 113–125.

Prieto, A.R. 1996. 'Late Quaternary vegetational and climatic changes in the pampa grassland of Argentina', *Quaternary Research* **45**, 73–88.

Przybylak, R. 1997. 'Spatial and temporal changes in extreme air temperatures in the Arctic over the period 1951–1990', *International Journal of Climatology* **17**, 615–634.

Pujos, M., Latouche, C. and Maillet, N. 1996. 'Late Quaternary paleoceanography of the French Guiana continental shelf – clay mineral evidence', *Oceanologica Acta* **19**, 477–487.

Rackham, O. 1986. *The History of the Countryside.* Dent, London.

Raisbeck, G.M., Yiou, F., Bourles, D., Lorius, C., Jouzel, J. and Barkov, N.I. 1987. 'Evidence for two intervals of enhanced [10]Be deposition in Antarctic ice during the last glacial period', *Nature* **326**, 273–277.

Rajagopalan G., Sukumar, R., Ramesh, R., Pant, R.K. and Rajagopalan, G. 1997. 'Late Quaternary vegetational and climatic changes from tropical peats in southern India – an extended record up to 40,000 years BP', *Current Science* **73**, 60–63.

Ram, M., Donarummo, J. and Sheridan, M. 1996. 'Volcanic ash from Iceland similar to 57,300 yr BP eruption found in GISP2 (Greenland) ice core', *Geophysical Research Letters* **23**, 3167–3169.

Rasmussen, T.L., VanWeering, T.C.E. and Labeyrie, L. 1997. 'Climatic instability, ice sheets and ocean dynamics at high northern latitudes during the last glacial period (58–10 Ka B.P.)', *Quaternary Science Reviews* **16**, 71–80.

Raymo, M. E. 1994. 'The initiation of northern hemisphere glaciation', *Annual Review of Earth and Planetary Sciences* **22**, 353–383.

Raymo, M. E. and Ruddiman, W. F. 1992. 'Tectonic forcing of late Cenozoic climate', *Nature* **359**, 117–122.

Raynaud, D., Chappellaz, J., Barnola, J.M., Korotkevich. Y.S. and Lorius, C. 1988. 'Climate and CH_4 change in the Vostok ice core', *Nature* **333**, 655–657.

Richard, P.J.H. 1994. 'Wisconsinan late-glacial environmental change in Quebec: a regional synthesis', *Journal of Quaternary Science* **9**, 165–170.

Richmond, G.M. and Fullerton, D.S. 1986. 'Summation of Quaternary glaciations in the United States of America', *Quaternary Science Reviews* **5**, 183–196.

Sabin, A.L. and Pisias, N.G. 1996. 'Sea-surface temperature changes in the northeastern Pacific Ocean during the past 20,000 years and their relationship to climate change in northwestern North America', *Quaternary Research* **46**, 48–61.

Schlüchter, C. 1986. 'The Quaternary glaciations of Switzerland, with special reference to the Northern Alpine Foreland', *Quaternary Science Reviews* **5**, 413–419.

Schneider, R.R., Muller, P.J. and Ruhland, G. 1995. 'Late Quaternary surface circulation in the east equatorial south Atlantic: evidence from alkenone sea surface temperatures', *Paleoceanography* **10**, 197–219.

Schweingruber, F.H. 1988. *Tree Rings. Basics and Applications of Dendrochronology*. Reidel, Dordrecht.

Scott, L., Steenkamp, M. and Beaumont, P.B. 1995. 'Palaeoenvironmental conditions in South Africa at the Pleistocene–Holocene transition', *Quaternary Science Reviews* **14**, 937–947.

Semple, E.C. 1911. 'Influences of geographic environment', in *Human Geography. An Essential Anthology*, J. Agnew, D.M. Livingstone and A. Rogers (eds) 1996. Blackwell, Oxford, 252–267.

Shackleton, N.J. and Opdyke, N.D. 1973. 'Oxygen isotope and palaeomagnetic stratigraphy of equatorial Pacific core V28-238: oxygen isotope temperatures and ice volume on a 10^5 and 10^6 year scale', *Quaternary Research* **3**, 39–55.

Shackleton, N.J., Imbrie, J. and Hall, M.A. 1983. 'Oxygen and carbon isotope record of East Pacific core V19-30: implications for the formation of deep water in the late Pleistocene North Atlantic', *Earth and Planetary Science Letters* **65**, 233–244.

Shackleton, N.J., Backmann, J., Zimmerman, H., Kent, D.V., Hall, M.A., Roberts, D.G., Schnitker, D., Baldauf, J.G., Desprairies, A., Homrighausen, R., Huddleston, P., Keene, J.B., Kaltenback, A.J., Krumsiek, K.A.O., Morton, A.C., Murray, J.W. and Westberg-Smith, J. 1984. 'Oxygen isotope calibration of the onset of ice-rafting and history of glaciation in the North Atlantic region', *Nature* **307**, 620–623.

Shane, P.A.R., Black, T.M., Alloway, B.V. and Westgate, J.A. 1996. 'Early to Middle Pleistocene tephrochronology of North Island, New Zealand – implications for volcanism, tectonism, and paleoenvironments', *Geological Society of America Bulletin* **108**, 915–925.

Shaw, P.A. and Thomas, D.S.G. 1996. 'The Quaternary paleoenvironmental history of the Kalahari of Southern Africa', *Journal of Arid Environments* **32**, 9–22.

Shemesh, A., Burckle, L.H. and Hays, J.D. 1995. 'Late Pleistocene oxygen isotope records of biogenic silica from the Atlantic sector of the Southern Ocean', *Paleoceanography* **10**, 179–196.

Siegert, M.J. 1997. 'Quantitative reconstructions of the last glaciation of the Barents

Sea: A review of ice-sheet modelling problems', *Progress in Physical Geography* **21**, 200–229.

Singh, B. 1997. 'Climate change in the greater and southern Caribbean', *International Journal of Climatology* **17**, 1093–1114.

Singh, G. and Geissler, E.A. 1985. 'Late Cainozoic history of fire, vegetation, lake levels and climate at Lake George, New South Wales, Australia', *Philosophical Transactions of the Royal Society of London* **B311**, 379–447.

Smith, G.W., Nance, R.D. and Genes, A.N. 1997. 'Quaternary glacial history of Mount Olympus, Greece', *Geological Society of America Bulletin* **109**, 809–824.

Spielhagen, R.F., Bonani, G., Eisenhauer, A., Frank, M., Frederichs, T., Kassens, H., Kubik, P.W., Mangini, A., Nørgaard-Pedersen, N., Nowaczyk, N.R., Schäper, S., Stein, R., Thiede, J., Tiedemann, R. and Wahsner, M. 1997. 'Arctic Ocean evidence for late Quaternary initiation of northern Eurasian ice sheets', *Geology* **25**, 783–786.

Stokes, S., Thomas, D.S.G. and Washington, R. 1997. 'Multiple episodes of aridity in southern Africa since the last interglacial period', *Nature* **388**, 154–158.

Stuiver, M. and Brazunias, T.F. 1993. 'Sun, ocean, climate and atmospheric $^{14}CO_2$: an evaluation of causal and spectral relationships', *The Holocene* **3**, 289–305.

Sun, D.H., Shaw, J., An, Z.S., Cheng, M.Y. and Yue, L.P. 1998. 'Magnetostratigraphy and palaeoclimatic interpretation of a continuous 7.2 Ma Late Cenozoic aeolian sediment from the Chinese Loess Plateau', *Geophysical Research Letters* **25**, 85–88.

Sundaram, R.M., Rakshit, P. and Pareek, S. 1996. 'Regional stratigraphy of Quaternary deposits in parts of Thar Desert, Rajasthan', *Journal of the Geological Society of India* **48**, 203–210.

Szeicz, J.M., MacDonald, G.M. and Duk-Rodkin, A. 1995. 'Late Quaternary vegetation history of the central Mackenzie Mountains, Northwest Territories, Canada', *Palaeogeography Palaeoclimatology Palaeoecology* **113**, 351–371.

Talma, A.S. and Vogel, J.C. 1992. 'Late Quaternary paleotemperatures derived from a speleothem from Cango Caves, Cape Province, South Africa', *Quaternary Research* **37**, 203–213.

Thomas, E., Booth, L., Maslin, M. and Shackleton, N.J. 1995. 'Northeastern Atlantic benthic foraminifera during the last 45,000 years – changes in productivity seen from the bottom up', *Paleoceanography* **10**, 545–562.

Thompson, L.G., Mosley-Thompson, E., Davis, M.E., Lin, P.-N., Henderson, K.A., Cole-Dai, J., Bolzan, J.F. and Liu, K.-B. 1995. 'Late glacial stage and Holocene tropical ice core records from Huascarán, Peru', *Science*, **269**, 46–50.

Thompson, L.G., Yao, T., Davis, M.E., Henderson, K.A., Mosley-Thompson, E., Lin, P.-N., Beer, J., Synal, H.-A., Cole-Dai, J. and Bolzan, J.F. 1997. 'Tropical climate instability: the last glacial cycle from a Qinghai–Tibetan ice core', *Science* **276**, 1821–1825.

Tzedakis, P.C. 1993. 'Long-term tree populations in northwest Greece, through multiple Quaternary climatic cycles', *Nature* **364**, 437–440.

Tzedakis, P.C. 1994. 'Vegetation change through glacial–interglacial cycles: a long pollen sequence perspective', *Philosophical Transactions of the Royal Society of London* **B354**, 403–432.

Unganai, L.S. 1997. 'Surface temperature variation over Zimbabwe between 1897 and 1993', *Theoretical and Applied Climatology* **56**, 89–101.

Van Andel, T.H. 1989. 'Late Quaternary sea-level changes and archaeology', *Antiquity* **63**, 733–745.

Vandenberghe, J., Zhisheng, A., Nugteren, G., Huayu, L. and Van Huissteden, K. 1997. 'New absolute time scale for the Quaternary climate in the Chinese loess region by grain-size analysis', *Geology* **25**, 35–38.

Vasari, Y., Glückert, G., Hicks, S., Hyvärinen, H., Simola, H. and Vuorela, I. 1996. 'Finland', in *Palaeoecological Events During the Last 15000 years. Regional Syntheses of Palaeoecological Studies of Lakes and Mires in Europe*, B.E. Berglund, H.J.B. Birks, M. Ralska-Jasiewiczowa and H.E. Wright (eds). John Wiley and Sons, Chichester, 281–351.

Versteegh, G. J. M. 1997. 'The onset of major Northern hemisphere glaciations and their impact on dinoflagellate cysts and acritarchs from the Singa section, Calabria (southern Italy) and DSDP Holes 607/607A (North Atlantic)', *Marine Micropaleontology* **30**, 319–343.

Villalba, R. 1994. 'Tree-ring and glacial evidence for the Medieval warm epoch and the Little Ice age in southern South America', *Climatic Change* **26**, 183–197.

Villalba, R., Boningsegna, J.A., Veblen, T.T., Schmelter, A. and Rubulis, S. 1997. 'Recent trends in tree-ring records from high elevation sites in the Andes of northern Patagonia', *Climatic Change* **36**, 425–454.

Walker, M.J.C., Bohncke, S.J.P., Coope, G.R., O'Connell, M., Usinger, H. and Verbruggen, C. 1994. 'The Devensian/Weichselian Late-glacial in northwest Europe (Ireland, Britain, north Belgium, The Netherlands, northwest Germany)', *Journal of Quaternary Science* **9**, 109–118.

Wang, P.X. and Sun, X.J. 1994. 'Last glacial maximum in China: comparison between land and sea', *Catena* **23**, 341–353.

Wanner, H., Pfister, C., Brazdil, R., Frich, P., Frydendahl, K., Jonsson, T., Kington, J., Lamb, H.H., Rosenorn, S. and Wishman, E. 1995. 'Wintertime European circulation patterns during the Late Maunder Minimum cooling period (1675–1704)', *Theoretical and Applied Climatology* **51**, 167–175.

Weaver, P.P.E. and Pujol, C. 1988. 'History of the last deglaciation in the Alboran Sea (western Mediterranean) and adjacent North Atlantic as revealed by coccolith floras', *Palaeogeography Palaeoclimatology Palaeoecology* **64**, 35–42.

Webb, P.N. and Harwood, D.N. 1991. 'Late Cenozoic glacial history of the Ross embayment, Antarctica', *Quaternary Science Reviews* **10**, 215–237.

Webb, T. III, Cushing, E.J. and Wright, H.E., Jr 1983. 'Holocene changes in the vegetation of the Midwest', in *Late-Quaternary Environments of the United States, Vol. 2, The Holocene*, H.E. Wright Jr (ed.). University of Minnesota Press, Minneapolis, 142–165.

West, R.G. 1980. 'Pleistocene forest history in East Anglia', *New Phytologist* **85**, 571–622.

Wheeler, D. 1995. 'Early instrumental weather data from Cadiz: a study of late eighteenth century and early nineteenth century records', *International Journal of Climatology* **15**, 801–810.

Williams, D.F., Thunell, R.C., Tappe, E., Rio, D. and Raffi, I. 1988. 'Chronology of the Pleistocene oxygen isotope record 0–1.88 m.y. BP', *Palaeogeography Palaeoclimatology Palaeoecology* **64**, 221–240.

Williams, M.A.J., Dunkerley, D.L., De Deckker, P., Kershaw, A.P. and Stokes, T. 1993. *Quaternary Environments*. Arnold, London 1st edn.

Williams, K.M., Andrews, J.T., Weiner, N.J. and Mudie, P.J. 1995. 'Late Quaternary paleoceanography of the mid-continental shelf', *Arctic and Alpine Research* **27**, 352–363.

Williams, MA.J., Dunkerley, D.L., De Deckker, P., Kershaw, A.P. and Stokes, T. 1998. *Quaternary Environments*. Arnold, London, 2nd edn.

Williams, P.W. 1996. 'A 230 Ka record of glacial and interglacial events from Aurora Cave, Fjordland, New Zealand', *New Zealand Journal of Geology and Geophysics* **39**, 225–241.

Worsley, T.R., Nance, D.R. and Moody, J.B. 1991. 'Tectonics, carbon, life, and climate for the last three billion years: a unified system', in *Scientists on Gaia*, S. H. Schneider and P.J. Boston (eds). MIT Press, Cambridge, Massachusetts, 200–210.

Xiao, J., Inouchi, Y., Kumai, H., Yoshikawa, S., Kondo, Y., Liu, T.S. and An, Z.S. 1997. 'Biogenic silica record in Lake Biwa of Central Japan over the past 145,000 years', *Quaternary Research* **47**, 277–283.

Yiou, F., Raisbeck, G.M., Bourles, D., Lorius, C. and Barkov, N.I. 1985. '^{10}Be in ice at Vostok Antarctica during the last climatic cycle', *Nature* **316**, 616–617.

Zagwijn, W.H. 1992. 'The beginning of the ice age in Europe and its subdivisions', *Quaternary Science Reviews* **11**, 583–591.

Zheng, B.X. and Rutter, N. 1998. 'On the problem of Quaternary glaciations, and the extent and patterns of Pleistocene ice cover in the Qinghai–Xizang (Tibet) Plateau', *Quaternary International* **45–6**, 109–122.

Zielinski, G.A. and Mershon, G.R. 1997. 'Palaeoenvironmental implications of the insoluble microparticle record in the GISP2 (Greenland) ice core during the rapidly changing climate of the Pleistocene–Holocene transition', *Geological Society of America Bulletin* **109**, 547–559.

Zielinski, G.A., Mayewski, P.A., Meeker, L.D., Whitlow, S. and Twickler, M. S. 1996. 'A 110,000-yr record of explosive volcanism from the GISP2 (Greenland) ice core', *Quaternary Research* **45**, 109–118.

Index

acidity 56–7
Adhémar, Joseph 6
aeolian sediments 82–4
aerosols, acidic 57
Africa, low latitudes of 148–50
age estimation 16, 80, 81, 85–6, 90, 167
agriculture 102, 165
Alaska 115, 117, 119
alerce, used for dating 93
algae 20, 40, 58, 76, 132
alkenones 37, 42–3, 172
Alps 4, 5, 124, 125, 128
aluminium in ice cores 55, 161
Amazonia 47, 155–7
Amerasian basin 107–9
Americas, low latitudes of 155–9
Andes 47, 137, 139, 157–8
angiosperms 24, 159, 172
Anglian glaciation 27, 67, 74, 82
Antarctic zone 51, 106, 120–1
anthropogenic change 2, 77, 79, 102, 159, 164
Archaean aeon 17, 22
archaeology 44, 93, 100, 103
Arctic zone 106–19
Argentina 139–40
art, as evidence for climate change 90, 100–1
ash, from volcanoes 58, 59, 142
Asia, low latitudes of 150–4
Asia, middle latitudes of 133
astronomical theory 7, 29–30, 60, 165, 169
Atherton Tableland, Australia 159, 160
atmosphere, composition of 1, 18–21, 23, 28, 51–2, 54, 165
Aurora Cave, New Zealand 85
Australia 71, 72, 142–4, 159–61
autotrophs 10, 42

Baoji loess deposit 83

barium in ocean sediments 46
Barkham soil 82
Barton County, Kansas, USA 134, 136
beech 71, 91, 113, 120, 140
beetles 6, 37, 112, 130
benthic organisms 38, 172
beryllium 56
biodiversity 19, 156
biogenic material 34, 172
biogeochemical cycles 1, 11, 15, 20, 172
biogeography 39, 44, 172
biosphere 1, 9, 15, 18
biostratigraphy 24
biota of the Earth 1, 11, 15, 22, 23, 172
birch 70, 74, 76, 78, 113, 114
bird populations 160
Blytt, work with Sernander 6
Britain: glacial environment of 67, 129–30; lakes in 74, 76; palaeoecology of 77–80
Brückner, Edward 3, 5, 125, 128
Buckland, William 6
Burundi 80, 149

cadmium in ocean sediments 46
calcium carbonate 21, 38, 42, 84–5
calcium in ocean sediments 46
Cambrian period 19, 20
Canada 76, 115, 117, 118, 137
Cango Caves, South Africa 141
Cape Collinson interglacial 117
carbon: biogeochemical cycle of 1, 11, 54, 55, 165; burial of 20; isotopes of 7, 25, 80–1, 91, 151, 167
carbon dioxide 18, 20–1, 40, 54, 61, 169
carbonate deposits 35, 84–5, 107
Carboniferous period 20, 23
catastrophism 3, 4, 172
caves: calcium carbonate in 84–5, 141, 143, 175; paintings in 100

charcoal 93, 140, 160
chernozem soils 133
cherry blossom 90, 97
Chile 44, 93–4, 139
China 82–4, 97–8, 131, 151, 153
chitin 45, 172
chronology from tree-ring data 91–2
chronostratigraphy 33, 144, 172
chrysophytes 76, 172
church records 98, 99
clay deposits 47, 74, 82
Clements, Frederick 3, 9
CLIMAP project 47–8
climate: global 11, 24, 165–6; historical 97–9, 101; *see also* cooling, global; warming, global
climate change: and carbon cycle 55; causes of 6–7, 28–31; evidence for 60–1, 65, 95, 131, 136, 141; prediction of 2, 168, 173; reaction of vegetation to 119, 157
climate cycles 69, 167
climax, climatic 9, 113
coal, formation of 20
coccolithophores 34, 42–4
Coleoptera 6, 37, 112, 130
Colombia 72, 155–6
Conger, Paul 6
Congo 150
continental drift 3, 11, 172; *see also* plate tectonics
cooling: in Europe 125; global 23, 29, 35–7, 40, 58, 93, 169; *see also* glacial stage
Coope, G. Russell 6
coral reef 28, 153, 154
cordillera 134, 173
Costa Rica 155, 158
Cretaceous period 22, 23
Crete 98–9
Croll, James 7
Cromerian interglacial 26, 82, 114
crust of the Earth 16
cybernetics 9
cycle of erosion 2, 3, 8, 173
cypress, used for dating 93

Darwin, Charles 4, 8; *see also* evolution
dating techniques 2, 7, 80–1, 85–6, 90, 167
Davis, William Morris 8
de Charpentier, Jean 4
de Geer, Gerard 3, 7
deforestation 10
deglaciation after last ice-age 121
dendrochronology 90–7, 173
deserts 85, 100, 149, 150–1

determinism 8–9, 173
detrivores 10
deuterium oxide 53
Devensian glacial advance 67
Devonian period 20
diatoms 6, 34, 40, 120
diluvial theory 3, 4
dinosaurs, extinction of 23
DMS (dimethyl sulphide) 58
Dome C ice core 51, 59
drought 90, 95, 99
DTR (Diurnal Temperature Range) 103
dust: in ice cores 55, 59, 61; *see also* loess
DVDI (Dry and very dry index) 98
dynamic equilibrium 10, 60, 173

Earth history 1, 3, 15–28
earthquakes 18
ecosystems 3, 9–10, 165, 173
Eemian interglacial 26, 75, 112–14, 117
elm 71, 76, 78, 91, 113
Elsterian ice advance 27, 68, 111, 124
environmental change: development of ideas 1–13; human-induced 77, 79, 102, 159, 164; *see also* climate change
environmental determinism 8–9, 173
equilibrium, dynamic 10, 60, 173
equinoxes, precession of 7, 29–30
erosion, cycle of 2, 3, 8, 173
erratics 3, 4, 107
eucalyptus 159
Eurasia: high latitudes of 113–15; middle latitudes of 131–3
Europe: climate of 99–100; middle latitudes of 124–31; palaeoecology of 68, 75, 77
eustasy 5, 153, 173
evolution 8, 15, 22, 149, 160, 165; *see also* Gaia hypothesis
extinctions 22–3, 166

famine 90
feedback 10, 29, 55, 60, 173
Fennoscandinavia 47, 66, 75
Ferdynandovian interglacial 26, 74
Finland 112, 113
flood(s) 29, 91, 97–8; the biblical 3, 4
flowering, dates of 90, 99
foraminifer 34–5, 38–9, 43, 112
Forbes, Edward 5
forcing factors 29, 30, 55, 173
forest boundary 69, 137–8, 150, 153, 154, 155–6
forests: in Antarctica 120; boreal 112, 113; in Britain

77–80; fragmentation of 157; in Greece 127; tropical 160, 170
fossils 6, 24, 112, 130; *see also* pollen record; ocean sediment record
fragmentation, of forests 157
France 71, 100, 129
Funza lake sequence 69, 72, 73, 155–6
Fynbos 141

Gaia hypothesis 4, 11, 15, 20, 166, 173
GCM (General Circulation Model) 2, 168, 173
Geikie, Archibald and James 3, 4
gelifluction 65
genetics 160
geographical cycle 2, 3, 8, 173
geoidal eustasy 153, 173
geological record: long-term 15–31; short-term 24–8, 165
geophysiology 20
germanium in ocean sediments 46
GISP ice cores 50, 51–3, 57, 59, 110, 131
glacial–interglacial cycles 155, 169
glacial-isostatic change in sea-level 5
glacial period, the last 79, 85, 129, 142
glacial stage (ice age) 2, 5, 22–9, 58, 61, 165, 173
glaciation: evidence for past 23, 24, 101; in high latitudes 106–21; in Himalayas 150–2; landforms of 65–7; in South America 137, 139; theory of 2, 3–5
glacio-eustatic change in sea-level 5
Gleason, Henry 9
gneiss 16, 173
Gondwana 12, 22
Great Salt Lake, USA 5, 134–5
Greece 71, 123, 127, 129
greenhouse effect 20, 22–4, 96; *see also* warming, global
greenhouse gases 29, 54, 111; *see also* carbon dioxide; methane
Greenland ice cores 51, 60, 106–11, 131
GRIP ice cores 50–3, 56, 58, 59, 61, 94
Grove, Gilbert 5
Günz glaciation 5, 27, 128
gyttja 74, 173

half-life 81, 86
halophyte 140, 173
heavy metals 55, 61
Heer, O. 6
Heinrich events 37, 174
heterotrophic organisms 174
Himalayas 150–2
history of Earth's surface 16–28
Hockham Mere, Britain 76, 78–9

holism 9, 10, 11
Holocene epoch 75, 78–9, 110, 168; in Britain 129–30; in North America 136; and Quaternary sequences 26; in South America 140; temperature reconstruction of 43, 45
hominid evolution 8, 149, 165, 174; *see also* evolution
Hoxnian interglacial 74
Huascarán ice core 69, 155, 157–8, 161
Hula Basin, Israel 71, 72, 123, 132, 133
human-induced environmental change 2, 77, 79, 102, 159, 164
Huon peninsula, New Guinea 150, 154
Hutton, John 3, 4
hydrosphere 1, 18, 28

ice: advances 47, 67, 96, 107, 124; ages (glacial stages) 2, 5, 22–9, 58, 61, 165, 173; cores 3, 7, 25, 50–62, 106–11, 157–8; sheets 59, 64, 69, 117, 120, 124
icebergs 3, 37, 47
Iceland Sea 108, 110
ice-rafted debris 174
Illinoian cold stage 26, 134
India 150, 151
Indonesia 150, 153, 154, 157
Industrial Revolution 102
insect fossils 6, 37, 112, 130
insolation 29
interglacial period 2, 4–5, 26–8, 42, 109, 117, 174; *see also* glacial stage
interpluvials 5
interstadial period 28, 52, 61, 142, 149, 174; *see also* glacial stage
Ioannina peat sequence 71, 123, 127, 129
Ireland 77, 91, 129
Irian-Jaya, Indonesia 150, 154, 157
iron 58
isostatic change in sea level 153, 174
isotope stages of oxygen 7, 33–8, 52–3, 60, 83, 94, 157–8, 174
isotopes 7, 81, 86
Israel 71, 72, 132, 133

Jamieson, Thomas 5
Japan 69, 71, 72, 132–3
jet streams 31, 174

Kalahari desert 149
Kennard, A.S. 6
Kenya 148, 149
Kurtak, Siberia 123, 133

Lake Baikal, Siberia 71, 73, 123, 132–3

Lake Biwa, Japan 69, 71, 72, 123, 132
Lake Bonneville, USA 134–5
Lake Elgennya, Siberia 115
Lake Eyre, Australia 161
Lake George, Australia 71, 72
Lake Hula, Israel 71, 72, 132, 133
Lake Lahontan, USA 134–5
Lake Lama, Siberia 114–16
Lake Owens, USA 71–3
Lake Pata, Amazonia 155, 156
Lake Praclaux, France 71
Lake Taupo, New Zealand 142, 144
lakes: maar 71, 174; pluvial 134; sediments from 67–77, 154; water levels in 134
land bridge 117, 161
landforms, glacial 65–7
larch 114, 115
Laurasia 12, 22
Laurentide ice sheet 66, 117
Libby, Willard 81
life as agent of change see Gaia hypothesis
life, origin of 15, 18–21
lithosphere 1, 9, 16, 18
Little Ice Age 59, 96–7, 101
lizard populations of Australia 160
loess 5, 24, 82–4, 131–4, 136, 151, 174
Lomonosov Ridge, Arctic 108–9
Lovelock, James 11; see also Gaia hypothesis
Lyell, Charles 4, 5
Lynch's Crater, Australia 159–60

maar lake 71, 174
Mackenzie Mountains, Canada 117–19
MacLaren, Charles 5
macrofossils from plants 6, 77
magnetic susceptibility 83
mantle of the Earth 16
marine environment: habitats 23; sediments 2, 7, 33–48, 52, 106, 107
Marks Tey, Britain 74
Mediterranean climate 123
meteorites 16, 23, 29
meteorology 90, 95–103
methane 29, 54, 61, 169
methanesulphonate 57–8
middens, packrat 85, 87
migration of flora and fauna 5–6, 87
Milankovitch theory 7, 29–30, 60, 165, 169
Mindel glaciation 5, 26, 127, 128
models of climate 2, 168, 173
molluscs 6, 44–5, 142
Mongolia 95, 96

monsoons 41, 46, 83
Moore Lake, Canada 76
moraines 65, 67, 128
Mount Kenya 148, 149
Mount Olympus 123, 127
mountain building 11, 15, 19, 22, 29
MSA (methanesulphonic acid) 57–8
muds: gyttja 74, 173; pelagic 107, 175
multiglaciation 6

nanoplankton 42, 174
natural selection 8
neodymium in ice cores 59
Netherlands 24, 125, 126
New Zealand 85, 142–5
newspaper accounts of weather 100
Nilgiris region, India 150, 151
nitrate in ice cores 56
nitrogen cycle 1
North America: high latitudes of 115–19; middle latitudes of 134–8; Wisconsin ice advance 68
Northwind Ridge, Amerasia basin 107–9
Norway 112–13

oak 70, 78, 91–3, 127, 133
ocean sediment record 2, 7, 33–48, 52, 106, 107
ODP (Ocean Drilling Programme) sites 148, 149, 159
open system 174
orbit of the Earth 6, 29–30
orogenesis 11, 15, 19, 22, 29, 174
ostracods 24, 45
Otiran glaciation 85
Owen, Richard 6
oxygen in the atmosphere 18, 21, 22
oxygen-isotope stages 7, 33–8, 52–3, 60, 83, 94, 157–8, 174

packrats 85, 87
paintings as evidence for climate change 90, 100–1
palaeo-catena 82
palaeoecology 6, 77
palaeosols 80–2, 83, 117, 118, 136
Palaeozoic era 17, 20
pampas 139–40
Pangaea 11–12, 19, 22
Papua New Guinea 150, 153, 154, 161
Patagonia 97, 139
peat 68, 70, 77, 127, 151
pedogenesis 127, 136, 175
pelagic muds 107, 175
Penck, Albrecht 3, 4, 125, 128
periglaciation 124, 132

permafrost 65, 69
Perraudin, Jean-Pierre 4
Peru 155, 157–8
Phanerozoic aeon 17, 22, 23
phosphorus in sediments 20, 22, 46
photographs as evidence for climate change 100–1
photosynthesis 20–1, 42
pine 78, 113, 127, 133; used for dating 70, 91–3, 94, 95
pingo 65
pioneer tree species 76
plankton 38, 42, 43, 76, 175
plate tectonics 11–13, 18, 22, 29, 153, 175
Playfair, John 4, 5
Pleistocene 6, 25, 141
Pliocene epoch 24, 25, 31, 115, 120
pluvial 5; lakes 134
polar regions 107–11, 120–1
pollen analysis 3, 6, 68–70, 166
pollen record: in Africa 148; in Americas 155–6; in Eurasia 132; in Europe 74–9, 125; in marine sediments 45–6; in Scandinavia 112; in South America 139
postglacial period 75–7
prairies 137–8
preboreal oscillation 61, 93–4
precession of the equinoxes 7, 29–30
precipitation 5, 56, 95, 150
primary producers 42, 175
productivity: 45, 73; indices of 46; of marine environment 40, 58
Proterozoic aeon 17, 22, 23
protozoa 38–9, 41
psammophyte 140, 175

Qinghai-Xizang 152; see also Tibet
Quaternary period 24–7, 35, 84, 137, 151, 153

radiocarbon dating 3, 7, 25, 80–1, 91, 167
radiolaria 34, 41
radiometric age determination 16, 80–1, 85–6, 175
rainfall 5, 90, 95, 99
records: historical 90, 97–101; religious 98, 99; weather 95, 99; see also geological record; ocean sediment record; pollen record
refugium 129, 156, 170, 175
Reid, Clement 6
reinforcing factors 29, 55, 175
religious records 98, 99
Riss glaciation 5, 26, 127, 128
Rodinia 19, 22
Ross Ice Shelf 121

Rusaka peat bog, Burundi 80
Rwanda 149

Saalian ice advance 26, 68, 111, 114, 124
Sahara desert 100, 149
salinity 41, 42, 45, 77
Sangamon interglacial 26, 117, 134
savanna 149, 156, 170
Scandinavia 3, 111–13
sclerophyll woodland 143, 159, 160, 175
sea-level changes: estimation of 39, 44, 114, 120; and ice sheets 19, 23, 28, 64, 117, 141, 151–4, 161
sediments: analysis of 6, 144; characteristics of 47; glacial 65–7, 109; from lakes 67–9; from oceans 2, 7, 33–48, 52, 106, 107
Sernander, work with Blytt 6
Shaitan glacial stage 114
Siberia 73, 95, 96, 107, 113–16, 132–3
silica 34, 41, 46, 73
silicate–carbonate buffering system 20–1
silt 82–4
smectite 47, 107
sodium in ice cores 55
soil, ancient 80–4, 117, 118, 136
South Africa 141
South America, middle latitudes of 137, 139
Spain 98, 102, 127
species divergence 160
speleothems 84–5, 141, 175
Spencer, Herbert 8
stadials 28, 29, 52, 175; see also interstadial period; glacial stage
steppe 71, 133, 140, 170
stratigraphy: alkenone 37, 43; oxygen isotope 7, 33–8, 52–3, 60, 83, 94, 157–8, 174
stress, environmental 91
stromatolites 20, 175
strontium in ice cores 45, 59
sulphur compounds, in ice cores 56
sulphur cycle 1, 57–8
Sumba Island, Indonesia 150, 153–4
Sun and solar energy 10, 29–30
supercontinents 12, 19, 22
survival of the fittest 8
Sweden 39, 93, 112, 113
systems theory 3, 4, 10, 11

Taillon glacier, France 101
Tajikistan 133
Tansley, Sir Arthur 3, 9, 157
Tasmania 142, 143
tectonics 11–13, 18, 22, 29, 153, 175

temperature: global 7, 18–24, 34–5, 51, 60–1; reconstruction of 110, 130–1, 137, 157; sea-surface 41, 42–4, 47–8, 113, 120; *see also* cooling; warming; glacial stage
Tenaghi-Philippon peat sequence 71, 123, 127, 129
tephra 143, 145
termination of cold stage 37, 42, 175
Tertiary period 24, 35
Thames river 101
Thar desert 150–1
thresholds in environmental systems 10, 164
Tibet 31, 61, 82, 152
Tierra del Fuego 139
till 65–6
Trade Winds 149
travertine 84–5, 175
tree-line 69, 155, 156
tree rings 90–7
tropical regions 147–61
troposphere 102, 175
tufa 84, 175
tundra 100, 110, 115, 117, 137

Uganda 149
uniformitarianism 4, 176
uranium-series dating 85–6
USA 72–3, 134–6, 138
Ussher (Archbishop of Armagh) 3

Valley Farm soil 82
varve chronology 3, 7, 66, 176
vegetation communities 9, 74, 119

Venetz, Ignace 4
Victoria, Australia 142–3
volcanoes 18, 56–7, 59–60, 95, 102, 142
von Bertalanffy, Ludwig 9
von Lozinski 5
von Post, Lennart 3, 6
von Richthofen, Ferdinand 5
Vostok ice core 3, 50, 51–8, 60

Wanganui Basin, New Zealand 142, 143, 145
warming: global 4, 23, 35, 62, 102–3, 166; in high latitudes 111; in Pliocene Antarctica 120
water, heavy 53
weather, records of 90, 95–103
weathering 21, 29
Wegener, Alfred 11
Weichselian glaciation 26, 68–9, 111, 114, 124, 139
Wilson cycles 22
wind-blown deposits 5, 24, 82–4, 132–4, 136, 151, 174
Wisconsin glaciation 68–9, 117, 134–5, 139
Wolstonian glacial advance 67, 74
wood, yielding tree-ring data 90–7
woodland, sclerophyll 143, 159, 160, 175
Wrey, Harold 7
Würm glaciation 5, 127, 128

xerophyte 140, 176
X-rays 92

Yangtze river 153
Younger Dryas 59, 80, 93–4, 110, 129, 136, 139, 157–8, 176